# Probability and Statistics: Theory and Exercises

Authored by

## Horimek Abderrahmane

*Mechanical Engineering Department*
*Ziane Achour University*
*Djelfa -Algeria*

**Probability and Statistics: Theory and Exercises**

Author: Horimek Abderrahmane

ISBN (Online): 978-981-5124-90-3

ISBN (Print):  978-981-5124-91-0

ISBN (Paperback): 978-981-5124-92-7

First published in 2024.

need for a court order if at any point you breach any terms of this License Agreement. In no event will any delay or failure by Bentham Science Publishers in enforcing your compliance with this License Agreement constitute a waiver of any of its rights.

3. You acknowledge that you have read this License Agreement, and agree to be bound by its terms and conditions. To the extent that any other terms and conditions presented on any website of Bentham Science Publishers conflict with, or are inconsistent with, the terms and conditions set out in this License Agreement, you acknowledge that the terms and conditions set out in this License Agreement shall prevail.

**Bentham Science Publishers Pte. Ltd.**
80 Robinson Road #02-00
Singapore 068898
Singapore
Email: subscriptions@benthamscience.net

**BENTHAM SCIENCE**

# CONTENTS

# FOREWORD

The book "Probability and Statistics: Theory and Exercises" by Dr. Horimek Abderrahman, is an excellent scientific contribution, written in a simple, clear English language. Presented in a practical, user-friendly, and relevant manner, this book will enable engineering and science students and teachers to understand the concepts and theories of probability and statistics and apply them in real-life situations. This is a real contribution to spreading the culture of statistical thinking among the Algerian higher education community and business leaders in the country.

Statistical thinking has been adopted by world-class organizations as a way of understanding our complex world by describing it in relatively simple figures that capture essential aspects of its structure and components. Statistical thinking, through probability and statistics, provides us with evidence about our past and present and allows us to discover insights into the future. Through statistics, evidence-based decision-making processes can be established, so higher product or service quality can be delivered to customers; hence contributing to an improved Quality of Life. Indeed, statistics have been used to predict future events, and foresight even the future of nations and citizens, hence contributing to build a sustainable future.

As statisticians, we always refer to the famous statement from H.G. Wells, "Statistical thinking will one day be as necessary for efficient citizenship as the ability to read and write.", I would rather say that this event is already happening, and we need to dig deeper in this science through this book and other reference textbooks.

It is expected that this manuscript will be a useful guide for all individuals who are involved in studying events and phenomena related to social, technical, engineering, or medical sciences.

**Mohamed Aichouni**
Professor of Quality Engineering and Management
University of Hail, Saudi Arabia

# PREFACE

We all know the importance of probability and statistical calculations in everyday life. The trend toward perfect precision requires not making errors; unfortunately, the error is human. Consequently; researchers are majorly interested in reducing this error, by means of the mastery of the world of chance which is really not as random as it seemed 500 years ago. Currently, we can anticipate events while knowing the past and the laws of evolution thanks to the laws of probability and data censuses. Therefore, we can learn the basics of probability and statistics has become indispensable to specialists in mathematics and economics, as well as to field engineers who have a large part of their duties based on statistical analyses.

This course is intended primarily for students of technical sciences at the undergraduate level; but it is also beneficial to those enrolled in master's program and even to those who have already graduated. Without a doubt, this course offers considerable help to the teachers, with numerous illustrative examples and exercises solved in detail with discussions.

**Horimek Abderrahmane**
Mechanical Engineering Department
Ziane Achour University
Djelfa -Algeria

## ACKNOWLEDGMENT

I would like to thank all those who have posted detailed lessons and exercises on the subject of "probability and statistics" on the internet. I would like to especially thank Professor **Mohamed AICHOUNI** who honored me by writing the Forward page after consulting this book despite his other commitments.

# PART 1: PROBABILITIES

## CHAPTER 1

# Reminder on the Theory of Sets

**Abstract:** Accidental events in chemical and process industries may have catastrophic consequences. The present chapter aims to discuss the hazards and risks in chemical and process industries, where chemical species are used and/or transformed. After defining the concept of chemical risk, the possible accidental events in the process industry are presented based on their probability of occurrence. Some examples of relevant chemical accidents that occurred in the past are thoroughly discussed further. Safety measures *(i.e.,* preventive and protective procedures) in safety and process industries and primary and secondary reactions are also described. Finally, a screening method capable of providing a hazard evaluation by calculating the power released during the thermal decomposition of a substance *(i.e.,* the CHETAH method) is presented.

**Keywords:** Accidents, Chetah method, Probability, Primary and secondary reactions, Risk.

## INTRODUCTION

The analysis of specialized books in probability (and statistics) shows that the majority of them begin with a passage, or even an entire chapter on the sets and their relationships, followed by the probability course itself. The idea is that, exactly, the same laws are applied with the difference of nomination, of course in the sense of probability. The probability is defined as the ratio between the number of favorable cases to that of possible ones. In the theory of sets, it is the ratio (or percentage) of a part of the set to the set itself, counting the number of elements in each of them. The same logic for all known laws in sets theory.

For this, we will start with a brief reminder of sets and their main laws, which will serve as a beneficial introduction to the course given the simplicity of manipulations of the laws on sets. Gradually, the definitions used in probability will be introduced in a flexible and clear manner.

## DEFINITION

A set is *a collection* of clearly defined objects called elements of that set [1-3].

**Examples:**

-The set of odd numbers less than 14;

-All the vowels of the alphabet;

-All the students in our class.

**Warning:** We cannot say for example: "the set of large numbers" or "the set of intelligent people"…etc, because it is not clear what would be the elements of these "sets" !.

**Notations:** The sets are denoted by letters, most often in upper case: *A, B, C, D*, ... The elements of the set are then *listed* in braces { }.

**Examples:**

*A*={0 ; 1 ; 2 ; 3 ; 4 ; 5 ; 6} ;

*B*={1 ; 3 ; 5 ; 7 ; 9 ; 11 ; 13} ;

*V*={a ; o ; u ; i ; e ; y}

These sets are said to be defined by ***enumeration*** or by ***extension***.

When the number of elements in a set is very large or even infinite, we *cannot* enumerate them all. To define such a set, we give a property of its elements which makes it possible to understand what these elements are: we then say that the set is defined in ***comprehension***.

**Examples:**

*M*={integers below 9000} ;

*M*={*x/x* is an integer less than 9000}

We read: "*M* is the set of elements *x* such that *x* is an integer less than 9000".

**Note:** Some sets can be defined by enumeration and comprehension.

**Example:**

*G*={*x/x* is an even number between 8 and 17}     or     *G*={8 ; 10 ; 12 ; 14 ; 16}

## Cardinality of a Set

We call the cardinal of a set $A$, the number of elements of this set.

## Examples

$A=\{8; 10; 12; 14; 16; 18; 20\}$

Card ($A$)=7

$B=\{2^{nd}$ year Mechanical Engineering students$\}$

Card ($B$)=68 (for example)

## The Symbols $\in$ and $\notin$

For any set $E$ and any element $x$:

- If $x$ is an element of $E$, we write $x \in E$ and we read: « *x belongs to E* »
- If $x$ is not an element of $E$, we write $x \notin E$ and we read: « *x does not belong to E*»

**Example:** IF $E=\{5; 9; 12\}$ we write : $5 \in E$, $9 \in E$ et $12 \in E$, but $8 \notin E$, $2 \notin E$,… *etc.*

## Equal Sets

Two sets are *equal* if they have the same elements.

**Example:** Which of the following sets are: equal?, Different ?

$A=\{1; 2; 3; 4\}$

$B=\{1; 4; 5; 7\}$                          $A \neq B$   /   $B \neq C$   /   A=C

$C=\{3; 1; 4; 2\}$

## Empty Set

If the set contains no elements. It is noted as $\varnothing$ .

**Example:**   $A=\{x/x$ is a student in our class who is over 3m tall$\}$;   So: $A=\varnothing$

## VENN Diagram

To represent a set, we draw a closed line called a VENN diagram and put the elements of the set inside this line, the others outside.

## Example:

We give the set $E$={c ; t ; r ; p} and the elements m; g; 45. The VENN diagram for this example is schematized as follows (Fig. **1.1**):

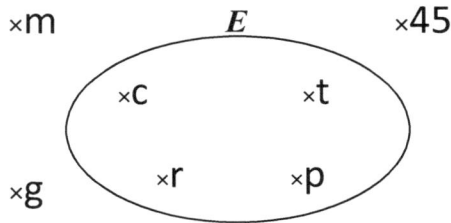

×m           *E*           ×45

×c           ×t

×g      ×r       ×p

**Fig. (1.1).** VENN diagram for a set.

To represent two sets on the same VENN diagram, it is necessary to provide a place for the elements which belong to both sets at the same time, for the elements which only belong to one of the two sets and for those which do not belong to neither of the two sets. Each element should only appear once on the diagram.

## Example:

We give the sets $A$={1; 4; 6} ; $B$={2; 3; 4; 5} and the elements 7 ; 8 and 9. VENN diagram for this example is as follows (Fig. **1.2**) :

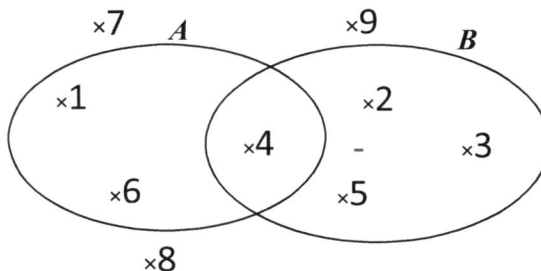

×7  *A*           ×9   *B*

×1           ×2

×4       -   ×3

×6       ×5

×8

**Fig. (1.2).** VENN diagram for two sets.

## Subsets

If all the elements of a set $E$ are also elements of a set $F$, we say that $E$ is a subset of $F$, or a part of $F$, or that $E$ is included in $F$.

## Notation:

-If $E$ is a subset of $F$ we write:

$$E \subset F \text{ and we read: } « E \text{ is included in } F »\qquad\qquad(1.1)$$

- If $E$ is not a subset of $F$ (*i.e.* if $E$ has at least one element that does not belong to $F$), we write:

$$E \not\subset F \text{ and we read: } « E \text{ is not included in } F »\qquad\qquad(1.2)$$

## Properties

- The empty set $\varnothing$ is included in any set $E$: $\varnothing \subset E$.
- Each set is included in itself: $E \subset E$.

**Warning** The symbols $\in$ and $\notin$ are used between an element and a set, while the symbols $\subset$ and $\not\subset$ are used between two sets.

## Example

Let the sets: $A=\{1; 3\}$; $B=\{1; 3; 5; 7\}$ and $S=\{1; 2; 3; 4; 5; 6; 7\}$.

$$A \subset B \text{ et } B \subset S$$

## Complementary of a Set

The complement of $A$ in $S$, corresponds to the set of elements which do not belong to $A$ but are in $S$. In logical writing: "element not $A$". This is the negation of $A$. We note it $A^c$ or $\overline{A}$ (Fig. **1.3**).

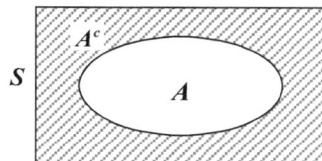

**Fig. (1.3).** Set A and its complimentary.

## Example

Let the sets: $A=\{1\;;3\}\;;$ $B=\{1\;;3\;;5\;;7\}$ and $S=\{1\;;2\;;3\;;4\;;5\;;6\;;7\}$

$$A^c=\{2\;;4\;;5\;;6\;;7\}\;;\qquad B^c=\{2\;;4\;;6\}$$

## Intersection, Union and Difference of two Sets

### *Intersection of two Sets*

Let $E$ and $F$ be two sets. The set of elements that belong to $E$ and $F$ is called *intersection* of $E$ and $F$ and is denoted by $E\cap F$ (Fig. **1.4**). Thus:

$$E\cap F=\{x/\,x\in E \text{ and } x\in F\} \tag{1.3}$$

If $E\cap F=\varnothing$ , we say that $E$ and $F$ are incompatible or mutually exclusive.

## Property:

$E\cap E^c=\varnothing$

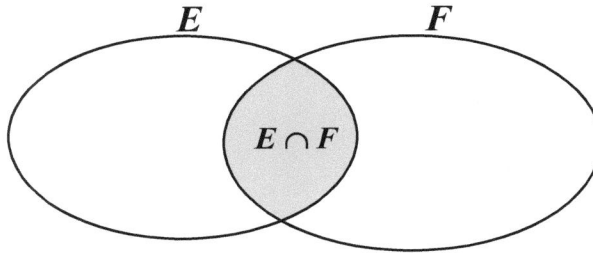

**Fig. (1.4).** Intersection of two sets.

### *Union of two Sets*

Let $E$ and $F$ be two sets. The set of elements that belong to $E$ or $F$ is called the union of $E$ and $F$ and is denoted by $E\cup F$ (Fig. **1.5**). Thus:

$$E\cup F=\{x/\,x\in E \text{ or } x\in F\} \tag{1.4}$$

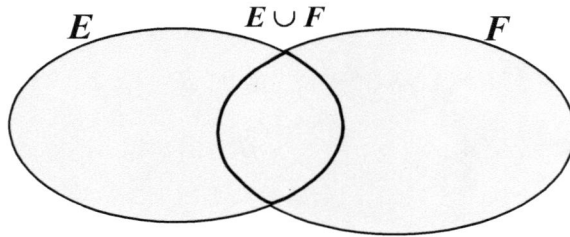

**Fig. (1.5).** Union of two sets.

## *Difference of two Sets*

Let $E$ and $F$ be two sets. The set of elements which belong to $E$ and which do not belong to $F$ is called the difference of $E$ and $F$ and is denoted by **E\F** (Fig. **1.6**).

$$E \setminus F = \{x/\, x \in E \text{ and } x \notin F\} \tag{1.5}$$

Likewise:

$$F \setminus E = \{x/\, x \in F \text{ and } x \notin E\} \tag{1.6}$$

In general, we can write:

$$A^c = S \setminus A \tag{1.7}$$

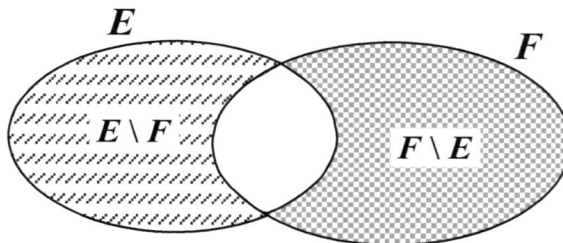

**Fig. (1.6).** Difference of two sets.

## **Property:**

-$E\backslash F$ and $F\backslash E$ have no common element, so:

$$E\backslash F \cap F\backslash E = \varnothing \tag{1.8}$$

## Main Properties of Sets

For *A* and *B,* two finite sets (Fig. **1.7**).

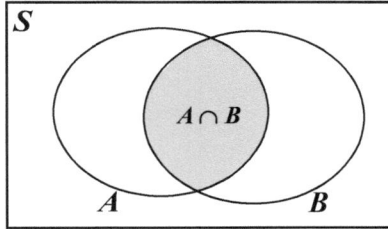

**Fig. (1.7).** Two sets in general situation.

We have:

$$Card(A \cup B) = Card(A) + Card(B) - Card(A \cap B) \qquad \textbf{(1.9)}$$

$$\begin{aligned} A \cup B &= B \cup A \\ A \cap B &= B \cap A \end{aligned} \qquad \text{..............(Commutative)} \qquad \textbf{(1.10)}$$

$$\begin{aligned} {A \cup B}^{\,c} &= A^c \cap B^c \\ {A \cap B}^{\,c} &= A^c \cup B^c \end{aligned} \qquad \text{..............(Two laws of De Morgan)} \qquad \textbf{(1.11)}$$

For three sets *A*, *B* and *C* (Fig. **1.8**)

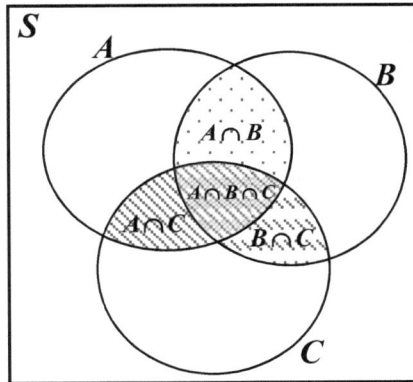

**Fig. (1.8).** Three sets in general situation.

We have:

$$Card(A \cup B \cup C) = Card(A) + Card(B) + Card(C)$$
$$- Card(A \cap B) - Card(A \cap C) - Card(B \cap C) \qquad (1.12)$$
$$+ Card(A \cap B \cap C)$$

$$A \cup B \cup C = A \cup B \cup C$$
$$A \cap B \cap C = A \cap B \cap C \qquad \dots\dots\dots\dots \text{(Associative)} \qquad (1.13)$$

$$A \cup B \cap C = A \cup B \cap A \cup C$$
$$A \cap B \cup C = A \cap B \cup A \cap C \qquad \dots\dots\dots\dots \text{(Distributive)} \qquad (1.14)$$

**Set of Parts of a Set (Powerset or Powerset)**

In mathematics, *the set of parts of a set* refers to the set of subsets of this set.

**Definition:** Let $E$ be a set. The set of parts of $E$, denoted generally $\wp(E)$, or $P(E)$, sometimes, is:

$$P(E) = \{A | A \subset E\} \qquad (1.15)$$

**Example:** Let be $E = a; b; c$ , a set of three elements. The set of parts of $E$ is:

$$P(E) = \left\{ \phi; \{a\}; \{b\}; \{c\}; \{a;b\}; \{a;c\}; \{b;c\}; \underbrace{\{a;b;c\}}_{E} \right\} \qquad (1.16)$$

*Cardinality*

Let be $E$ a finite of $n$ elements. Then, the set $P(E)$ of parts of $E$ is finite, and has $2^n$ elements.

$$Card(P(E)) = 2^{Card(E)} = 2^3 = 8 \qquad (1.17)$$

## The Tribe

**Definition:** Let $\Omega$ be a set. A family $F$ of parts of $\Omega$ is called a tribe if it satisfies the following properties:

i.   $\Omega$ is an element of $F$;
ii.  If $A$ is an element of $F$, then $A^c$ is an element of $F$, (stability by complement);
iii. If $(A_i)_{i \in N}$ are elements of $F$, then $\bigcup_{i \in N} A_i$ is an element of $F$. (denumerable union stability).

<u>**Warning:**</u> A tribe on $\Omega$ is a set whose elements are parts of the set $F$.

**Note:** For the set $\Omega$, the $P(\Omega)$ set of parts of $\Omega$ is *a tribe*. In the case where $\Omega$ is finite or countable, *we will always take as a tribe* on $F, P(\Omega)$. On the other hand, this choice is impossible when one considers a larger universe (that is to say not countable) as $\mathcal{R}$.

## Properties

***i.***  $\varnothing$ is an element of $F$ ;

If $A_{i \, i \in N}$ are elements of $F$, then $\bigcap_{i \in N} A_i$ is an element of $F$. (denumerable intersection stability).

# CHAPTER 2

# Introduction to Basic Definitions in Probability

**Abstract:** In this chapter, the basic definitions of probability theory are presented. The logic of presenting the results as events and their quantification, in order to know those that are highly probable or the reverse, are all detailed. It should be noted that in this chapter, we are only interested in knowing how to determine the value of the probability for a single random experiment.

**Keywords:** Events, Probability definition, Probability space, Random experiment.

## INTRODUCTION

In this chapter, we will present many definitions essential in the calculation of the probability. As already mentioned, calculating a probability comes to calculating the percentage of achievement or non-achievement of a phenomenon (or a result). So, if we determine the total space of possibilities and assume (even without doing) the phenomenon, it remains to determine its results (elements for a set) which can be classified according to our interest (characteristic of the set). The assumption of the phenomenon is called random experience (total set or universe) to be determined, the outcomes of which are called events (elements or subsets).

## RANDOM EXPERIMENT (RE)

**Definition:** A Random Experiment is a renewable experiment, the result of which cannot be predicted, and which, when repeated under identical conditions, does not necessarily give the same result each time it is repeated [3]. Each renewal of the experiment is called a trial (test, throw, *etc.*). A trial can combine several elementary trials, either consecutively or simultaneously. Trials can be more or less independent. The set of possible outcomes, of a random experiment constitutes the base of this experiment.

**Example:** If you toss a coin, you cannot predict which side it will land on. So: *Throwing a coin is a random experience in the sense of probability.*

## Events

**Definition:** Let $\Omega$ be a set with a tribe $F = P(\Omega)$. The elements of $F$ are called events or issues.

**Horimek Abderrahmane**

## Elementary Events

Let *E* be a *random experiment*, an event *A* linked to the experiment *E*, is said to be an elementary event if it is only realized by a single outcome of this experiment.

**Example:** We roll a dice (Fig. **2.1**). The exits are one of the 6 numbers appearing on the upper face after immobilizing the dice. The event "*The number on the top side of the die is 4*" is an elementary event.

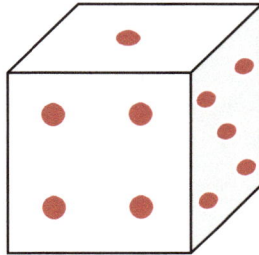

**Fig. (2.1).** A six-sided dice.

**Note:** An event can be a set of elementary events, and can be seen as a subset of $\Omega$, the set of all possible elementary events called *fundamental space* or *universe*.

**Example:** For a dice roll: $\Omega = \{1 ; 2 ; 3 ; 4 ; 5 ; 6\} = \{1;2;3;4;5;6\}$

## Certain Event

Let *E* be a random experiment and *A* an event linked to this random experiment. We say that *A* is a *certain event*, if it is *realized whatever the outcome of the experiment E*.

**Example:** We roll a dice. The event: "the number on the top side of the dice is less than 7", is a certain event.

**Note:** $\Omega$ is a certain event and $\varnothing$ is an empty event.

## Impossible Event

Let *E* be a random experiment and *A* is an event. We say that *A* is an *impossible event*, if *it is not realized, whatever the outcome of the experiment E*.

**Example:** We roll a dice. The event: "the number on the top side of the dice is number 7" is an impossible event.

## *Complementary Event*

Let $E$ be a random experiment and $A$ is an event. We say that $\bar{A}$ (or $A^c$) is the complementary event of $A$, if it is composed of elementary events not realized by $A$. It is defined as:

$$\bar{A} = \Omega - A \tag{2.1}$$

**Example:** In the experiment of rolling a dice, if $A$ is the event: "have an even number on the upper face", then: $A = \{1; 3; 5\}$. The complementary event $\bar{A}$ (not $A$) is: "not to have an even number". Therefore: $\bar{A} = \{2; 4; 6\}$

**Properties:**

i. 
$$\bar{\bar{A}} = A \ ;$$

ii. 
$$A \cap \bar{A} = \varnothing \ ;$$

iii. 
$$A \cup \bar{A} = \Omega \ ;$$

iv. 
$$\bar{\Omega} = \varnothing \ ;$$

v. 
$$\bar{\varnothing} = \Omega \ . \tag{2.2}$$

## *Incompatible Events*

Let $E$ be a random experiment and two events $A$ and $B$ are linked to this random experiment. We say that they are *incompatible* if they *cannot be carried out simultaneously* (Fig. **2.2**).

**Fig. (2.2).** Two incompatible events.

**Example:** We roll a dice. The events: "*the number on the upper face is equal to* 2" and "*the number on the upper face is equal to* 5", are *incompatible* (Disjoint $A \cap B = \varnothing$ ).

**Note:** If $\overline{A}$ is the complementary event of $A$. These two events are incompatible.

### *Independent Events*

Let $E$ be a random experiment and two events $A$ and $B$ are related to this random experiment. We say that they *are independent* if *the realization of $A$ does not influence the realization of $B$* and *vice versa*.

### **Examples:**

- We roll a dice the first time, we get number 5, then we make a second throw, we get number 2. The events $A = 5 \ ; B = 2$ are independent;
- A vehicle manufacturing plant manufactures cars, where it is reported that there are cars that have a brake problem, and there are cars that have a speed recovery problem. The two events (braking problem and speed recovery) are independent.

**Notes:**

- If the realization of an event is influenced by that of the other, they become dependent;
- *Independent* and *incompatible* events should not be confused. In case of independent events, the occurrence of one of the events does not prevent that of the second.

**Example:** We roll two dice simultaneously. We denote by $A$, $B$, and $C$ the following events:

- I get 1 with the first dice (event $A$);
- I get 4 with the second dice (event $B$);
- I get 3 with the first dice (event $C$).

Events $A$ and $B$ are *independent* while events $A$ and $C$ are *incompatible*.

## DEFINITION OF A PROBABILITY

The passage from a description of sets to a quantification of random phenomena is done by means of a *probability measure* or simply *probability* denoted by $p$ [3-4].

**Definition:** We call probability on the fundamental space $\Omega$, endowed with a tribe $F$, a function $(p(F))$ towards the positive closed interval [0,1] ( $p: p(F) \rightarrow [0\,;1]$ ) , satisfying the two properties:

$$p(\Omega) = 1 \,(\text{Maximum probability}) \qquad (2.3)$$

$$\text{If } A \text{ and } B \text{ are incompatible: } p(A \cup B) = p(A) + p(B) \qquad (2.4)$$

**Notes:**

1. The properties *i.* and *ii.* are axioms. We have decided that a probability by definition satisfies these two properties. When we say "let $p$ be a probability ...", these two properties are automatically satisfied, we do not have to prove them.
2. Why were these two properties chosen for the definition of probability? First, let's notice that they are quite intuitive:

- $p(\Omega) = 1$, intuitively means that $\Omega$ well describes all the possible outcomes of the random experiment, which one cannot fall outside of $\Omega$ ;
- As for the second property, let us think of the case of two disjoint events $A$ and $B$: it seems natural that the probability has an additivity property of the type: $p(A \cup B) = p(A) + p(B)$. However, the additivity for two disjoint parts is not enough to give the *σ-additive*, we therefore preferred to take it as an axiom *σ-additive*.

**Properties:** For all events $A$ and $B$,

**Probability Space (Probabilized Space)**

Let $\Omega$ be a set provided with a tribe $F$. A probability $p$ on $(\Omega, F)$ is a function of $F$ in $\left[0,1\right]$ satisfying the following properties:

**i.**     $p(\bar{A}) = 1 - p(A)$ ;

**ii.**    $p(\varnothing) = 0$ ;

**iii.**   If : $A \subseteq B$ then : $p(A) \leq p(B)$ ;                                                    **(2.5)**

**iv.**    $0 \leq p(A) \leq 1$ ;

**v.**     $p(A \cup B) = p(A) + p(B) - p(A \cap B)$

*i.*If $\underset{i \in N}{A_i}$ are elements of $F$ two-by-two disjoint, then:

$$i. \quad p(\Omega) = 1 ; \tag{2.6}$$

$$p\left(\underset{i \in N}{\cup} A_i\right) = \sum_{i=1}^{n} p(A_i) \qquad (\sigma\text{-additive}) \tag{2.7}$$

The triplet $\Omega, F, p$ is then called a probability space.

**Remark 01:** We can generalize to realize the notions of union and intersection to $n$ events $A_1$; …; $A_n$ two by two disjoint.

$$p\left(\overset{n}{\underset{i=1}{\cup}} A_i\right) = \sum_{i=1}^{n} p \; A_i \tag{2.8}$$

**Mutually Independent Events**

Let the events be $A_1$; …; $A_n$. These events are said to be mutually independent for the probability $p$ if any subset of two or more of these events satisfies the following property: *"The probability of the conjunction of the events of the subsets is the product of their probabilities"*.

$$p\left(\underset{1 \leq j \leq k}{\cap} A_{ij}\right) = \underset{1 \leq j \leq k}{\prod} p \; A_{ij} \tag{2.9}$$

This expression is only correct if $A_{ij}$ are mutually independent with $k \geq 2$ for any family of two-by-two distinct elements.

**Remark 02:** If $\Omega$ is a fundamental space of $n$ disjoint events $A_1$; …; $A_n$. We have:

$$\bigcup_{i=1}^{n} A_i = \Omega \quad \Rightarrow \quad p\left(\bigcup_{i=1}^{n} A_i\right) = p\ \Omega\ = 1 \tag{2.10}$$

So, if the $A_1$; …; $A_n$ events are elementary events with the same probabilities, we say there is equiprobability. This probability is then worth **1/n**. And, in this case, the probability of an event **A** containing **k** elementary events is **k/n**.

More generally, we write:

$$p(A) = \frac{Card(A)}{Card(\Omega)} \tag{2.11}$$

Where $Card\ \ A$ denotes the number of elements of $A$.

We then find the well-known definition (Definition of LAPLACE).

$$\boxed{p(A) = \frac{N^{br}\ of\ favorable\ cases}{N^{br}\ of\ possible\ cases}} \tag{2.12}$$

**Example:** For the random experiment consisting of rolling a non-rigged dice: $\Omega =\ \ 1;2;3;4;5;6$

$$A_1 = \{1\};\ A_2 = \{2\};\ A_3 = \{3\};\ A_4 = \{4\};\ A_5 = \{5\};\ A_6 = \{6\}$$

$$p(A_1) = p_1 = p_2 = p_3 = p_4 = p_5 = p_6 = \frac{1}{6}$$

$$So;\ if\ \ A = \{1;3;5\}\ \Rightarrow\ p(A) = p(A_1) + p(A_3) + p(A_5) = \frac{3}{6}$$

*But how to calculate the number of favorable and possible cases for all situations?.*

There are four (04) cases, which can be summarized by the  following  diagram (Fig. **2.3**):

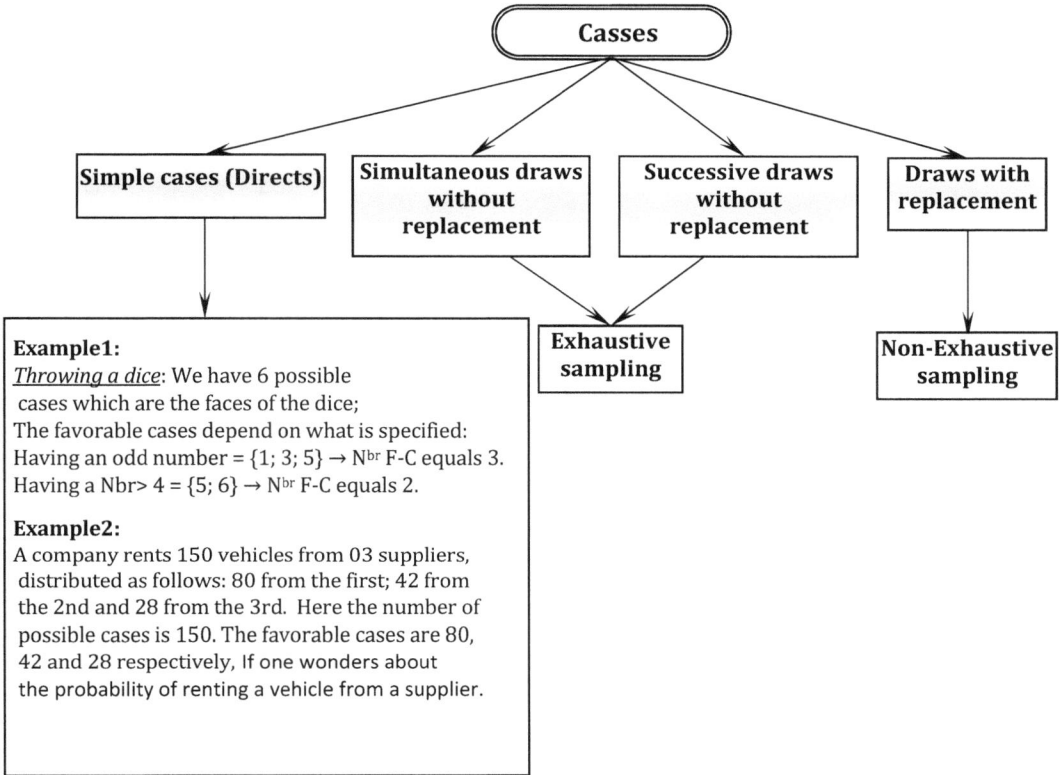

**Fig. (2.3).** Types of "favorable and possible cases".

The direct cases are very simple and do not pose a problem during the determination of the probability. The other difficult cases have been the subject of the chapter called combinatorial analysis, where laws have been demonstrated, which help to easily obtain the number of favorable and/or possible cases once the situation is very well understood. These counting techniques will be the subject of the next section.

# CHAPTER 3

# Combinatory Analysis

**Abstract:** This chapter presents in detail the calculation techniques for random experiments with a high number of outcomes that are generally impossible to calculate classically. The outcomes can be the number of favorable cases or that of possible ones, necessary to calculate the probability value. It is enough to imagine the situation well, to know which law should be used.

**Keywords:** Arrangements, Combinations, Permutations, Principle of multiplication.

## INTRODUCTION

The objective of combinatorial analysis (also known as Combinatorics or Counting Technique), is to learn to count the finite number of elements with large cardinalities.

## MULTIPLICATION PRINCIPLE (GENERAL PRINCIPLE OF ENUMERATION)

It allows counting the number of results of experiments which can be broken down into a series (succession) of sub-experiments [1,2,5].

**Principle:** Assuming that an experiment is the succession of $m$ sub-experiments. If the $i^{th}$ experiment has $n_i$ possible results, for $i = 1, 2,...., m$. So, the total number of possible outcomes of the overall experiment is:

$$N = \prod_{i=1}^{m} n_i = n_1 \times n_2 \times ...... \times n_m \qquad (3.1)$$

**Example 01:** You buy a suitcase with a code formed by 04 digits (from 0 to 9).

- How many ways can you choose your code?

**Solution:** We can imagine the code to be formed as a random experience, in which, the choice of each digit constructs a sub-experience.

The code can take many formulations, for example:

**0000 / 0001 / 0002 /……….. / 1111 / 1112 / ………. / 9997 / 9998 / 9999**

Therefore:

- The **first** digit can be chosen in **10 ways** (from **0** to **9**);
- The **second** digit can be chosen in **10 ways** (from **0** to **9**);
- The **third** digit can be chosen in **10 ways** (from **0** to **9**);
- The **fourth** digit can be chosen in **10 ways** (from **0** to **9**).

The total number of possible codes is therefore: **10×10×10×10=10⁴**     (10000 codes!!!).

The multiplication principle verifies this result. The number of codes is *N* while the number of possibilities of each digit is $n_i$. The code is made up of 04 digits, so *m*=4.

$$N = \prod_{i=1}^{4} digt_i = digt_1 \times digt_2 \times digt_3 \times digt_4 = 10 \times 10 \times 10 \times 10 = 10^4 \qquad (3.2)$$

**Example 02:** How many license plates can we have, if it is made up of: **02** different letters followed by **03** digits (from **0** to **9**)? The first digit (*on the left*) cannot be a **0**.

**Solution:** With the specified considerations, the plate has the following shape (examples):

**AB100 / AB101 / AB102 /……….. / BA100 / BA101 / ………. ZY998 / ZY999**

- The **first letter** can be chosen in **26 ways** (from **A** to **Z**);
- The **second letter** can be chosen in **25 ways** (from **A** to **Z** *except* the **first letter** chosen);
- The **first digit** can be chosen in **9 ways** (from **1** to **9**);
- The **second digit** can be chosen in **10 ways** (from **0** to **9**);
- The **third digit** can be chosen in **10 ways** (from **0** to **9**).

So the plates' number for our considerations is *N*. *m* equal to 5 (plate with 5 characters) and $n_i$ changes depending on the character:

$$N = \prod_{i=1}^{4} charac_i = lett_1 \times lett_2 \times digt_1 \times digt_2 \times digt_3 = 26 \times 25 \times 9 \times 10 \times 10 = 585000 \qquad (3.3)$$

**Important note:** All that we will see in what follows are only special cases of the principle of multiplication. We must always try to make the connection in order to be able to resolve very complex cases.

## PERMUTATIONS

There are two cases, permutations without repetition and those with repetition [1-2].

### Permutation without Repetition

A permutation without repetition of *n* distinct (different) elements $e_1, e_2, ...., e_n$, is an ordered rearrangement (the order is respected) without repetition of one or more elements of the *n* elements.

**Example:** We have 02 books. The 1st of Statistics (**S**) and the 2nd of Mathematics (**M**). We want to put them on a shelf.

- How many ways can we do this?

**Solution:** We will schematize the situation to better understand (Fig. **3.1**):

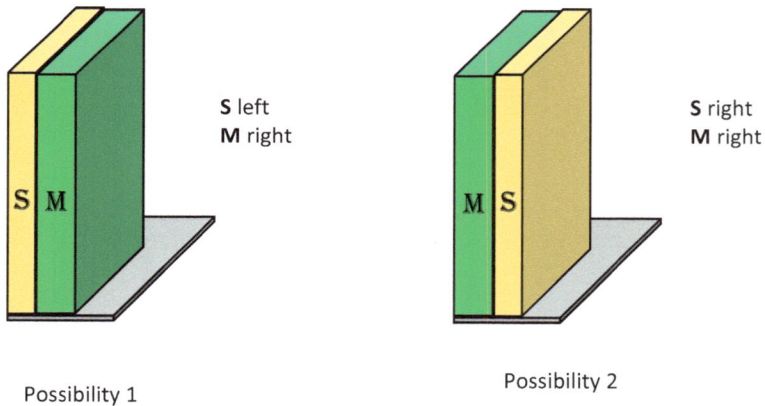

Possibility 1          Possibility 2

**Fig. (3.1).** Permutations for two elements.

For the two books we have two possibilities. Either the **S** on the left (starting direction) then **M** must necessarily be on the right, or the opposite. In reality, the space occupied by the two books is the same; we only swapped the positions (positions' permutation) between them.

We can also diagram the situation, using what we call "the tree diagram of probabilities", which we will see in detail later (Fig. **3.2**).

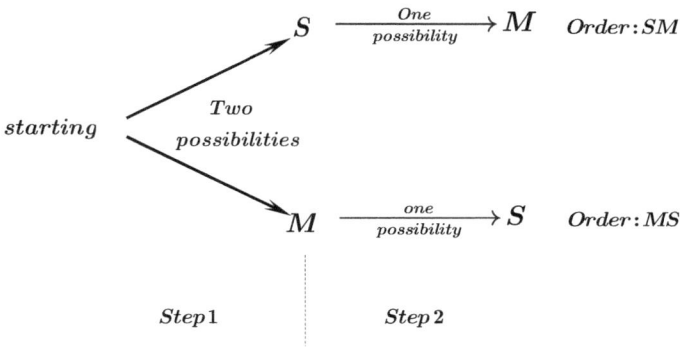

**Fig. (3.2).** Probability tree of permutations for two elements.

It is clear that there are two possibilities in total for this situation.

The *principle of multiplication* verifies this result. $N$ represents the possibilities' number. $m$ represents the number of steps (of books here) and $n_i$ the number of possibilities for each step.

$$N = 2 \times 1 = 2! = P_2 \qquad (3.4)$$

**Case of 03 books:** We add an Economics book ($E$) to the set of two books ($S$ and $E$). How many classifications are possible this time (Fig. **3.3**)?

With the tree diagram (Fig. **3.4**):

It is clear that if the number $n$ of elements increases, the number of possibilities will always be this number in factorial ($n!$). In addition, the number of possibilities is none other than the number of possible permutations between these elements regardless of their number.

In general (can be proved by Mathematical induction), we write:

$$\boxed{P_n = n!} \qquad (3.5)$$

And we read permutation of $n$ elements ($P$ of $n$).

We therefore have six (06) possible ways.

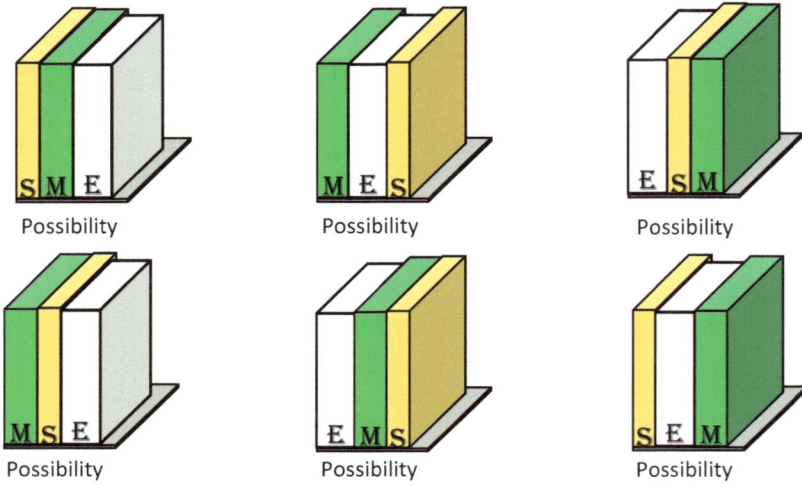

**Fig. (3.3).** Permutations for three elements.

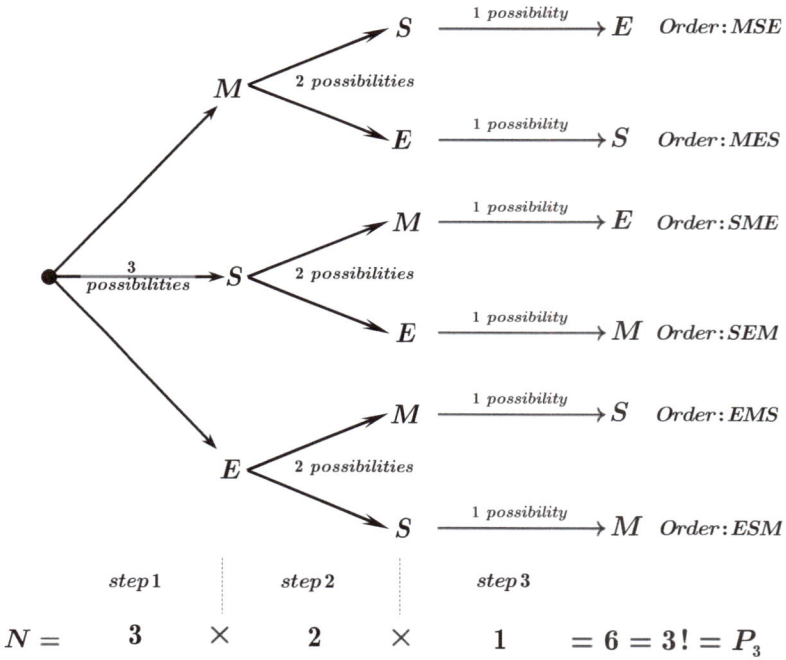

**Fig. (3.4).** Probability tree of permutations for three elements.

## Permutation with Repetition

When the elements to be classified in order contain **k** identical elements (impossible to distinguish between these **k** elements), the number of permutations between the **k** elements must be subtracted (the whole order must be calculated only once) from the total number of possibilities.

**Example 1:** How many possible words (anagrams: with or without meaning) we can have from the letters of the word "SUCCESSFUL".

**Solution:** Here are some examples of its anagrams:

**SUCCESSFUL  /  USCCESSFUL  /  SCUCESSFUL  /  CCESSFULSU  /**
…………….

- The number of permutations if the letters were different is: $P_{10} = 10!$
- But, one cannot distinguish between the three **S** nor the two **C** or the two **U**. Therefore, the cases where one makes permutations between the **S(s)** only, the **C(s)** only, or the **U(s)** only. Two of them (**S** and **C** for example), or all of them at the same time, without permuting the other letters will give the same word. This word must be counted only once ;

*Example*:

**SUCCESSFUL ≡ SUCCESSFUL ≡ SUCCESSFUL**

**≡ SUCCESSFUL ≡ SUCCESSFUL ≡ SUCCESSFUL**

- The number of permutations:

between the **S(s)** is $P_3 = 3!$ / between the **C(s)** is $P_2 = 2!$ / between the **U(s)** is $P_2 = 2!$

- So, the number of distinct permutations, actually carried out is:

$$P = \frac{10!}{3! \times 2! \times 2!} = 151200 \qquad (10! = 3628800 = 24\ times) \qquad (3.6)$$

Generally, we write:

$$P_n^{n_1,n_2,\dots,n_k} = \frac{n!}{n_1! \times n_2! \times \dots \times n_k!}$$  (3.7)

## ARRANGEMENTS

An arrangement is a permutation of **k** elements taken from **n** *distinct* elements (**k≤n**). There are also two cases: Without repetition or replacement and with repetition [1, 5].

### Arrangement without Replacement

Consider the set $\Omega = \{A; B; C; D\}$. Their elementary sets (with a single element) are four $\{A\}$, $\{B\}$, $\{C\}$ *and* $\{D\}$. So to form sets (called arrangements) with two elements (**k**=2), we have 12 ways as shown on the tree diagram (Fig. **3.5**).

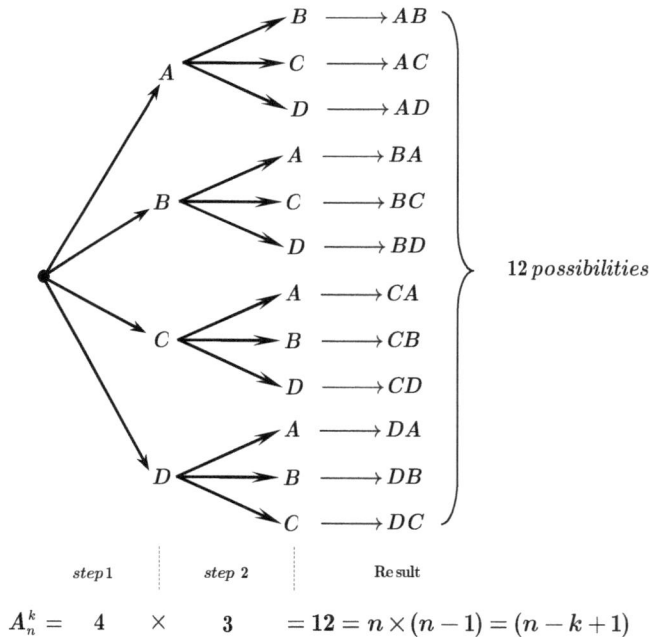

$$A_n^k = \quad 4 \quad \times \quad 3 \quad = 12 = n \times (n-1) = (n-k+1)$$

**Fig. (3.5).** Arrangements without replacement of two of four elements.

From the tree, we can see that the sets formed by two elements differ:

- Either by the order between the elements: $AB \neq BA$ ;

- Either by the nature of an element or more: $AB \neq AD$   /// $AB \neq CD$ ;

- Either by nature and order at the same time: $AB \neq CA$

So, in a no-replacement arrangement, the **order** and the **nature** of the elements must be respected.

As mentioned under the tree, the number of possibilities is named $A_n^k$. We read "Arrangement of **k** elements among **n**".

In this example: **k**=2 and **n**=4.

The first element has **04** *possibilities* to be chosen. Once the first element is chosen, the universe of choices becomes of cardinality **less than 1** compared to the initial state. The second element, therefore, has only **03** *possibilities*. In total, there are **12 possibilities** according to the multiplication principle.

We can see that *step* 2 has **(-1)** possibilities compared to *step* 1. In addition, the stopping of the draw is done at exactly **(n-k +1)** of possibilities for this *step* (=**3** *in the example*).

If now we draw **3** *times* in succession and *instead of two*, it is clear that we are going to have, **4** *possibilities* in *step* 1, **3** *possibilities* in *step* 2, and **2** *possibilities* in *step* 3. This makes a total of **24 possibilities**.

For this case: **k=3**   et   **n=4**.

And:

$$A_n^k = A_4^3 = 4 \times 3 \times 2 = 24 = n \times (n-1) \times (n-2) = n \times (n-1) \times (n-k+1) \text{ (3.8)}$$

So we stop here too at $(n-k+1)$.

We can easily imagine that always (for any value of **n** and **k≤n**), that the stop occurs at $(n-k+1)$. In general, we have:

$$A_n^k = n \times (n-1) \times (n-2) \times \ldots \ldots \times (n-k+1) \tag{3.9}$$

This expression can be compacted by multiplying and dividing by $(n-k)!$. And we come to the more popular expression:

$$A_n^k = \frac{n!}{(n-k)!} \tag{3.10}$$

## Example

- How many words with 03 distinct letters can we form from the Alphabet of 26 letters?
- How many 03-letter words can we make in a 26-letter Alphabet?
- What is the probability of having a word with 03 distinct letters?

**Solution:** We can imagine the Alphabet as a set of **$n$** different elements (**$n$**=26) from which we will draw 03 elements in succession (**$k$**=03). You can draw with the replacement of the drawn element (possible repetition) or without replacement (no possible repetition). It is the details that fix the case.

- For words with **03 *distinct*** letters:

The difference between the letters implies the *non-possibility* of repetition. This implies that we are in the case of a draw without replacement in order to avoid the possibility of drawing the letter two or three times. This is exactly the case with an arrangement without replacement:

$$n_{word\,3L\neq} = A_{26}^3 = \frac{26!}{(26-3)!} = \frac{26!}{23!} = 26 \times 25 \times 24 = 15600$$

- For words with **03 letters**: without any precision for the letters. Here we do not exclude the possibility of repeating the letter.

Using the principle of multiplication: $n_{word3L} = 26 \times 26 \times 26 = 26^3 = 17576$

- For the ***probability*** of having a word with **03 distinct** letters:  $p = \dfrac{n_{word\,3L\neq}}{n_{word\,3L}} = \dfrac{15600}{17576} \simeq 0.88757$

## Arrangement with Replacement

We keep the same previous example with the set $\Omega = \{A; B; C; D\}$. We are going to form sets with two elements ($k$=2) but before choosing the 2$^{nd}$ element we put back the 1$^{st}$ drawn. We have 16 ways as shown in the tree below (Fig. **3.6**).

It is very clear that returning the drawn element to the original set will renew the draw experience. Therefore, we will reproduce the experience and automatically we will have the same number of possibilities. For $k$ draws, we will have: $\underbrace{n \times n... \times n}_{k\,steps} \cdot$

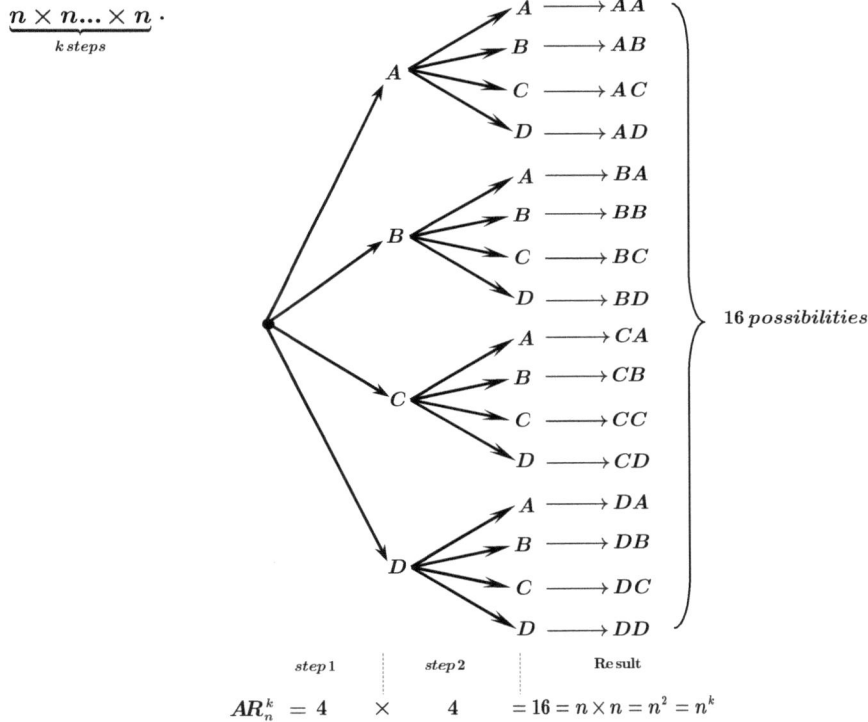

Fig. (**3.6**). Arrangements with replacement of two of four elements.

Finally, we write:

$$\boxed{AR_n^k = n^k} \qquad (3.11)$$

The second case of the **3**-letter words previous example, is a case that can be treated as an arrangement with replacement.

## Properties

-   $A_n^k \leq AR_n^k$ : Arrangements without repetitions are included in those with repetitions (cases among them);

-   For $k = n$ :   $A_n^k = \dfrac{n!}{(n-k)!} = \dfrac{n!}{(n-n)!} = \dfrac{n!}{0!} = n! = P_n$

-   For $k = 1$ :   $A_n^k = \dfrac{n!}{(n-1)!} = n = n^1 = AR_n^k$

## COMBINATIONS

In a simple way, the combination is an arrangement of **k** elements taken from **n** distinct elements (**k≤n**) where the **order** between the elements is **no longer important**. We are only interested in the elements constituting the formed subset. There are also two cases, without replacement (without repetition) and with replacement [1-2].

### Combination without Replacement

We will use the case presented for the arrangement to clarify the difference but for $\Omega = $   $A; B; C$   for more clarity and space saving (Fig. **3.7**).

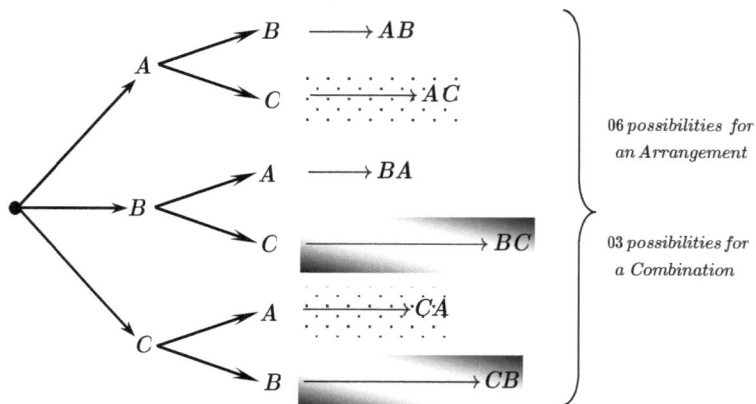

**Fig. (3.7).** Combinations without replacement of two of three elements.

From the tree diagram, if the problem was for *an arrangement*, then:

$$AB \neq BA \ // \ AC \neq CA \ // \ BC \neq CB$$

But when the problem is for a combination, the order is not important, and thus:

$$AB \equiv BA \; // \; AC \equiv CA \; // \; BC \equiv CB$$

So for a combination, we are only interested in the elements.

The only possibility for the order to become unimportant in a draw, is to make the draw of the $k$ elements simultaneously. That is, we draw all the $k$ elements at once. So, if you have a *simultaneous draw* in an experiment, you have to work directly with the combination.

The analysis of the difference between an arrangement and a combination leads to subtracting the number of possible permutations between the $k$ elements, to go from the number of possibilities for the arrangement to that for the combination (as made for permutation with repetition).

Consequently, we obtain:

$$C_n^k = \frac{n!}{k! \times (n-k)!} \tag{3.12}$$

### Example

A $2^{nd}$ year mechanical engineering class contains 60 students. We want to form groups of 07 students, to organize for them a visit to a factory (one group for a section).

- How many ways can we choose the students for the groups (or how many possible groups)?

### Solution

It is clear from the details, that any 07 students can form a group regardless of their identities. In here, the order does not matter and the number of possibilities is exactly that of the combinations of 7 out of 60.

$$C_{60}^7 = \frac{60!}{7! \times (60-7)!} = \frac{60 \times 59 \times 58 \times 57 \times 56 \times 55 \times 54}{7 \times 6 \times 5 \times 4 \times 3 \times 2 \times 1} = 386206920$$

- Now; if we have fixed tasks for the 07 students (for example: chief; assistant ...), the order becomes important (we will have precisely one student for a well specified task), and the number of possibilities becomes that of an arrangement.

## Properties

- $$C_n^k = C_n^{n-k}$$

*Proof* :

$$C_n^k = \frac{n!}{k!(n-k)!} = \frac{n!}{(n-n+k)!(n-k)!} = \frac{n!}{(n-(n-k))!(n-k)!} = \frac{n!}{(n-k)!(n-(n-k))!} = C_n^{n-k}$$

- $$C_n^k = C_{n-1}^k + C_{n-1}^{k-1}$$

*Proof* : $C_n^k = \dfrac{n!}{k!(n-k)!} = \dfrac{n(n-1)!}{k!(n-k)!} = \dfrac{(n-k+k)(n-1)!}{k!(n-k)!}$

$$= \frac{(n-k)(n-1)!}{k!(n-k)!} + \frac{k(n-1)!}{k!(n-k)!} = \frac{(n-k)(n-1)!}{k!(n-k)(n-k-1)!} + \frac{k(n-1)!}{k(k-1)!(n-k)!}$$

$$= \frac{(n-1)!}{k!((n-1)-k)!} + \frac{(n-1)!}{(k-1)!(n-k)!} = \frac{(n-1)!}{k!((n-1)-k)!} + \frac{(n-1)!}{(k-1)!((n-1)-(k-1))!}$$

$$= C_{n-1}^k + C_{n-1}^{k-1}$$

The second property can be schematized as follows:

$$\boxed{C_{n-1}^{k-1}} + \boxed{C_{n-1}^{k}}$$
$$=$$
$$\boxed{C_n^k}$$

The variation of *n* and *k* leads to obtaining a very popular triangle of coefficients, known as ***Pascal's triangle*** (a very interesting triangle which deserves to be consulted in the literature). The calculation procedure is illustrated in the table below (Table **3.1**).

## Combination with Replacement

It is a very rare situation. Usually presented as a *logical flow* of the Combinatorial Analysis course only [5, 6].

**Table. 3.1. Triangle of Pascal.**

| $k \rightarrow$ | 0 | 1 | 2 | 3 | 4 | .... |
|---|---|---|---|---|---|---|
| $n_1\downarrow$ | 1 | 1 | | | | |
| 2 | 1 | 2 | 1 | | | |
| 3 | 1 | 3 | 3 | 1 | | |
| 4 | 1 | 4 | 6 | 4 | 1 | |
| 5 | 1 | 5 | 10 | 10 | 5 | 1 |
| . | . | . | . | . | . | . |

Assuming the set $\Omega = \ A;B;C;D\ $ , from which we want to form sets of two elements without considering the order, and the possibility of repeating the element. As shown in the tree diagram (Fig. **3.8**), we have a decrease of 1 in each possible case going down, given the non-importance of the order.

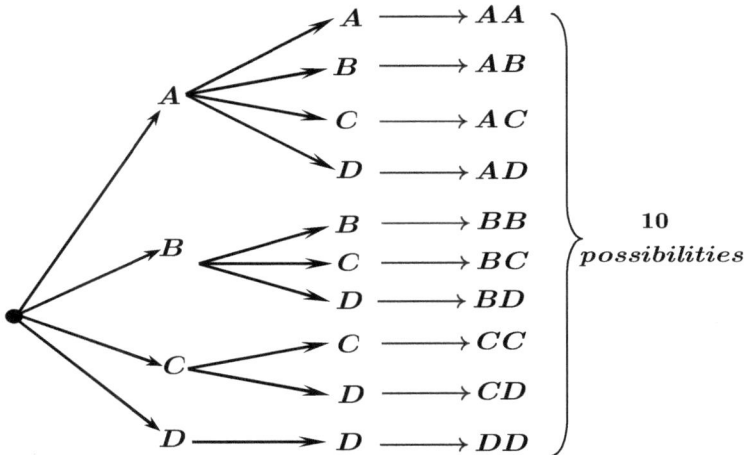

**Fig. (3.8).** Combinations with replacement of two of four elements.

We can show that the number of possibilities is:

$$K_n^k = C_{n+k-1}^k = \frac{(n+k-1)!}{k! \times (n-k)!}$$ (3.13)

## Property

- $K_n^k = K_{n+1}^k + K_{n-1}^{k-1}$

This property can be schematized as follows:

$$\boxed{K_{n-1}^{\,k}}$$
$$+ \;\; =$$
$$\boxed{K_n^{\,k-1}} \quad \boxed{K_n^{\,k}}$$

From which we can draw the following table (Table **3.2**):

**Table. 3.2. Schematization of the property.**

| $k \rightarrow$ <br><br> $n \downarrow$ | 0 | 1 | 2 | 3 | .... |
|---|---|---|---|---|---|
| 1 | 1 | 1 | 1 | 1 | .... |
| 2 | 1 | 2 | 3 | 4 | .... |
| 3 | 1 | 3 | 6 | 10 | .... |
| 4 | 1 | 4 | 10 | 20 | .... |
| . | . | . | . | . | . |

# CHAPTER 4

# Calculation of Probabilities

**Abstract:** After learning how to determine the probability value for a single random experiment, this chapter gives you the tools for compound random experiments. The probability tree that graphically presents the experiment is defined first, followed by numerous details and laws, leading to the mastery of calculation tricks for an experiment composed of n sub-experiments.

**Keywords:** Bayes' theorem, Conditional probability, Global probability, Total probability law, Tree diagram of probabilities.

## INTRODUCTION

After having defined in detail what is a probability (Laplace's law), and how to calculate it for simple cases and those with big cardinality, in this chapter, we move on to describing its calculation for complex cases with exploitation of the definitions seen on the events.

## PROBABILITY TREE

Let consider a random experience of rolling a non-rigged dice. Experience' outcomes can be represented by the following tree diagram of probabilities (Fig. **4.1**) [1]:

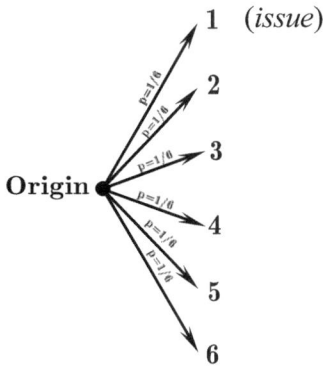

In this case, the dice is not rigged. So, all faces have the same chance of appearing. *i.e.* that we have equiprobability. The probability for such an issue is:

$$p = 1/n = 1/6 \ (n = 6 : \textbf{\textit{Nbr of faces}})$$

**Fig (4.1).** Probability tree for rolling a fair six-sided dice.

## Compatible Events

Two events $A$ and $B$ are compatible, if the realization of $A$ does not interfere (does not prevent) the realization of $B$ and *vice versa*.

## Example

$A$= « Have an **odd** number on the six-sided dice» $\Rightarrow A = \{1; 3; 5\}$ and $p(A) = 3/6 = 1/2$

$B$= « Have a **multiple** of **3** the six-sided dice» $\Rightarrow B = \{3; 6\}$ and $p(B) = 2/6 = 1/3$

For the event, $A \cup B = \{1; 3; 5; 6\}$ the probability will be: $p(A \cup B) = 4/6 = 2/3$

We can easily see that issue **3** exists in both events $A$ and $B$, so they are compatible. In addition, the probability of their union amounts to summing their probabilities independently and then subtracting that of the common outcome to be counted only once (see Chapter 2).

So, for every two compatible events, we have:

$$\boxed{p(A \cup B) = p(A) + p(B) - p(A \cap B)} \tag{4.1}$$

## Incompatible Events

Two events $A$ and $B$ are incompatible, if their simultaneous realization is impossible.

## Example

$A$= « Have an **odd** number on the six-sided dice» $\Rightarrow A = \{1; 3; 5\}$ and $p(A) = 3/6 = 1/2$

$B$= « Have a **multiple** of **2** the six-sided dice» $\Rightarrow B = \{2; 4; 6\}$ and $p(B) = 3/6 = 1/2$

For the event $A \cup B = \{1; 2; 3; 4; 5; 6\}$ the probability will be: $p(A \cup B) = 6/6 = 1$

Given the incompatibility of events, there are no common issues. The probability of union is therefore the direct summation of their probabilities.

We have: $A \cap B = \varnothing \implies p(A \cap B) = 0$

Thus, we have the law:

$$\boxed{p(A \cup B) = p(A) + p(B)} \tag{4.2}$$

**Note:** The laws of probability calculus for two compatible or incompatible events can be generalized to *n* events.

Let's take back our first example (throwing the dice). The weighted tree diagram (Fig. **4.2**) can be put in the following general form (for any considered event):

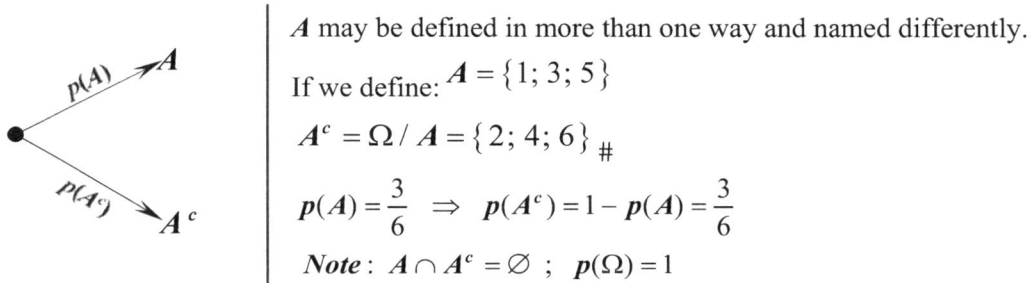

*A* may be defined in more than one way and named differently.

If we define: $A = \{1; 3; 5\}$

$A^c = \Omega / A = \{2; 4; 6\}_{\#}$

$p(A) = \dfrac{3}{6} \implies p(A^c) = 1 - p(A) = \dfrac{3}{6}$

*Note* : $A \cap A^c = \varnothing$ ; $p(\Omega) = 1$

**Fig. (4.2).** Generalized form of the probability tree for a RE.

If we roll the dice twice in succession, we can define the events $A = \{1; 3; 5\}$ on the 1st launch and $B = \{3; 6\}$ on the 2nd launch. The probability tree diagram will therefore be composed of two small trees, whose origin of the second is the end of the first regardless of the outcome. For this, we obtain the following tree diagram (Fig. **4.3**):

## CALCULATION RULE

Starting from a node, we realize the partition of a sub-universe. Here, for example, the roll of the 2nd dice is a sub-universe of the random experience made up of the two rolls [1-2].

**Rule 01:** The sum of the probabilities of all branches starting from the same node equals **1**.

$$p(A) + p(A^c) = \frac{1}{2} + \frac{1}{2} = 1 \quad //// \quad p(B) + p(B^c) = \frac{1}{3} + \frac{2}{3} = 1$$

**Rule 02:** The probability of a path is equal to the product of the probabilities written on its branches.

$$p(B \cap A) = p(B) \times p(A) = \frac{1}{3} \times \frac{1}{2} = \frac{1}{6} \quad //// \quad p(B^c \cap A^c) = p(B^c) \times p(A^c) = \frac{2}{3} \times \frac{1}{2} = \frac{1}{3}$$

**Note:** This rule (number 02) can be understood by means of the multiplication principle and Laplace's definition of a probability of a compound experiment:

$$p(B \cap A) = \frac{Nbr\ of\ Fav\,Cas\,(B) \times Nbr\ of\ Fav\,Cas\,(A)}{Nbr\ of\ Poss\,Cas\,(B) \times Nbr\ of\ Poss\,Cas\,(A)} =$$

$$\frac{Nbr\ of\ Fav\,Cas\,(B)}{Nbr\ of\ Poss\,Cas\,(B)} \times \frac{Nbr\ of\ Fav\,Cas\,(A)}{Nbr\ of\ Poss\,Cas\,(A)} = p(B) \times p(A)$$

(4.3)

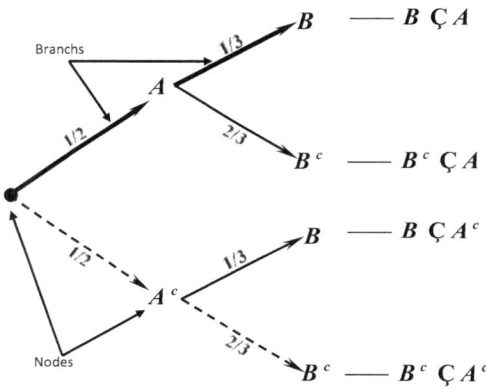

The event: **Odd** number on the 1st launch and *multiple* of 3 on the 2nd is described by the path with ***thick arrows***. It is written under the symbolization: $B\ Ç\ A$

It should be noted that for <u>several launches</u>, we start with the last result ($B$ here) then we move towards the beginning, launch by launch. This presentation indicates that the random experiment (the total experiment) is over.

The line composed by the two branches with discontinuous arrows ($B^c\ Ç\ A^c$) indicates that we have obtained a *non-multiple* number of 3 after obtaining an *even* number.

The same logic for the other lines.

**Fig. (4.3).** Generalized form of the probability tree for a combined RE.

So, if we move horizontally, we multiply the probabilities and we talk about a path, whereas when we move vertically (down), we sum them and we talk about branches.

## TOTAL PROBABILITY LAW

We take the previous figure here (Fig. **4.4**) [1-2].

- What is the probability of having a *multiple* of 3 on the 2nd launch (event **B**)?

The finding of **B** is possible by two paths: $B \cap A$    and   $B \cap A^c$

So:    $p(B) = p(B \cap A) + p(B \cap A^c)$

$= \dfrac{1}{3} \times \dfrac{1}{2} + \dfrac{1}{3} \times \dfrac{1}{2}$     (We have applied the two rules 01 and 02)

$= \dfrac{1}{3}$

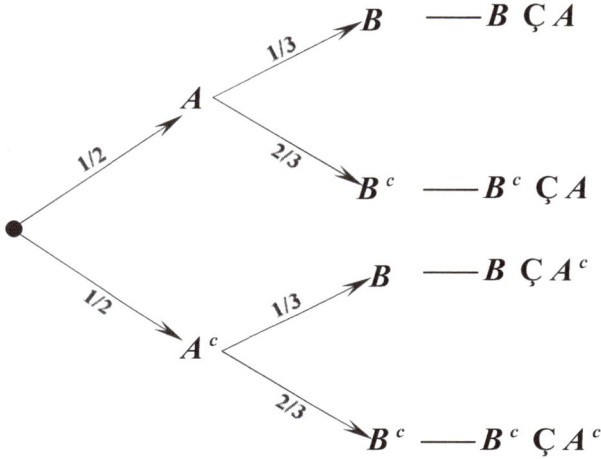

**Fig. (4.4).** Probability tree of the treated case.

## CONDITIONAL PROBABILITY

**Example:** Let an urn contain **03** *Red* balls and **02** *Green* balls (Fig. **4.5**) [1, 7]:

-    We draw two balls at random. We have two cases:

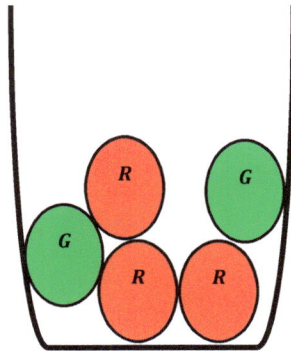

**Fig. (4.5).** Urn with 3 red balls and 2 green ones.

**Case 01:** The first ball drawn is put back before drawing the second (successive drawing with replacement). With this situation, we obtain the following tree diagram of probabilities (Fig. **4.6**):

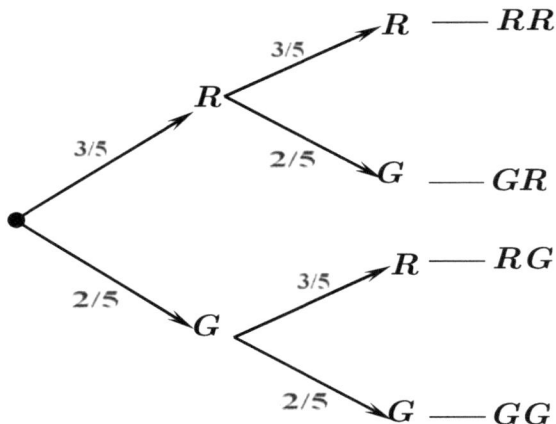

**Fig (4.6).** Probability tree for a draw with replacement.

The analysis of the probabilities of the branches shows that there is no change in their values between the first and the second draw. The cause is the return of the ball to the urn before the second draw is made, which leads to a reproduction of the experience. Here, we say that the second draw no longer depends on the first, and that the event in the second is ***independent*** of the first (for example having **R** after having **R** or **G** did not influence its probability).

We calculate:

$$p(G) = p(G \cap R) + p(G \cap G) = \frac{2}{5} \times \frac{3}{5} + \frac{2}{5} \times \frac{2}{5} = \frac{2}{5}$$

$$p_R(G) = \frac{2}{5} \quad (We\ read: p(G)\ given\ R)$$

$$p_G(G) = \frac{2}{5} \quad (We\ read: p(G)\ given\ G)$$

$$\boxed{p(G) = p_R(G) = p_G(G)} \qquad (4.4)$$

**Case 02:** The ball is not returned to the urn, and the second is drawn (successive draw without replacement). The probability tree diagram becomes: (Fig. **4.7**).

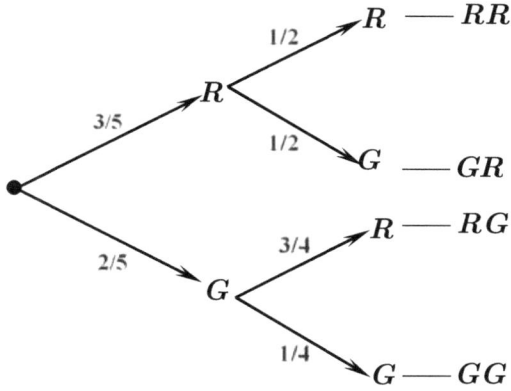

**Fig (4.7).** Probability tree for a draw without replacement.

We can clearly see the influence of the first draw on the probabilities of events of the second draw. The ball drawn in the first draw (not returned) changed the universe of samples. Here, the probability of the $2^{nd}$ draw event depends on the $1^{st}$ draw. The two events (of the $1^{st}$ and $2^{nd}$ draws) are therefore ***dependent***.

We calculate:

$$p(G) = p(G \cap R) + p(G \cap G) = \frac{1}{2} \times \frac{3}{5} + \frac{1}{4} \times \frac{2}{5} = \frac{2}{5}$$

$$p_R(G) = \frac{1}{2} \quad (We \ read: p(G) \ given \ R)$$

$$p_G(G) = \frac{1}{4} \quad (We \ read: p(G) \ given \ G)$$

$$\boxed{p(G) \neq p_R(G) \neq p_G(G)} \tag{4.5}$$

**Notes:**

- For ***dependent events***, care must be taken when determining the probabilities of the branches;

- For ***independent events***, it suffices to determine well for the first sub-experiment;

- The writing $p_R(G)$ (or: $p_G(G)$ ; $p_G(R)$ ; $p_R(R)$ ), means that it is the probability of the event **G** after having (or realization of) the event **R**. Its value comes only after the node after obtaining $\mathbf{R}$. We speak about the change of the universe of the random experiment; we call it "*conditional probability*". It is therefore clear that for *independent events*, this definition *does not matter*. While it is *very important* in the *second case*.

## TREE DIAGRAM OF PROBABILITIES IN THE GENERAL CASE (TWO RANDOM SUB-EXPERIMENTS)

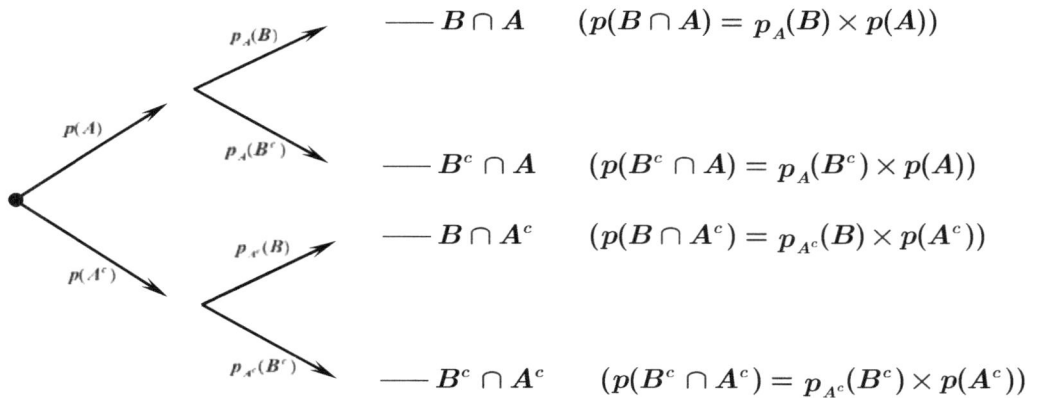

$$\begin{array}{ll}
p_A(B) \nearrow \quad\; — B \cap A & (p(B \cap A) = p_A(B) \times p(A)) \\[2em]
p_A(B^c) \searrow \quad — B^c \cap A & (p(B^c \cap A) = p_A(B^c) \times p(A)) \\[1em]
p_{A^c}(B) \nearrow \quad — B \cap A^c & (p(B \cap A^c) = p_{A^c}(B) \times p(A^c)) \\[2em]
p_{A^c}(B') \searrow \quad — B^c \cap A^c & (p(B^c \cap A^c) = p_{A^c}(B^c) \times p(A^c))
\end{array}$$

**Fig. (4.8).** Generalized probability tree for two random sub-experiments.

Thus [1, 2, 7]:

$$p(B) = p(B \cap A) + p(B \cap A^c) = p_A(B) \times p(A) + p_{A^c}(B) \times p(A^c) \qquad (4.6)$$

For $A$ and $B$ independent:

$$p(B \cap A) = p(B) \times p(A)$$

$$p(B \cap A^c) = p(B) \times p(A^c) \qquad (4.7)$$

This result can be extended to $n$ events $A_i$, two by two independent, each defined on a random sub-experiment (seen in Chapter 2):

$$\boxed{p\left(\bigcap_{i=1}^{n} A_i\right) = \prod_{i=1}^{n} p(A_i)} \qquad (4.8)$$

## GLOBAL (OVERALL) PROBABILITY

Let the universe $\Omega$ be made up of $B_1, B_2, ....., B_n$ incompatible events. Let $A$ be an event linked to the realization of one or more of the events $B_i$ $(i = 1 \rightarrow n)$. The (so-called global) probability of occurrence of event $A$ is calculated by the law [1, 7]:

$$p(A) = \sum_{i=1}^{n} p(B_i) \times p_{B_i}(A) \qquad (4.9)$$

In this expression, when $A$ is not related to one or more $B_i$ events, the value $p_{B_i}(A)$ will be *zero*. Furthermore; we can notice that this law is a generalization of the laws seen for the case of two launches, with only two incompatible events at the first launch (first sub-experience), whereas here they are of number $n$.

**Example:** Three identical boxes from the outside. The 1st contains **05** white balls (*W*) and **05** black balls (*K*). The 2nd contains **08** white balls and **02** black balls. The 3rd contains **04** white and **06** black (Fig. **4.9**).

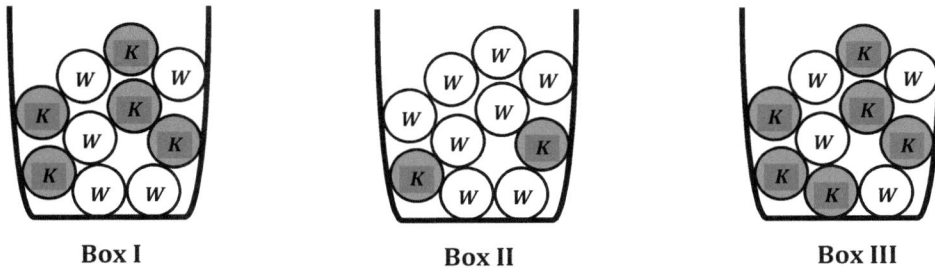

| Box I | Box II | Box III |

**Fig. (4.9).** Three boxes of 10 balls with different contents.

- We randomly draw a box, and then, from this, we draw a ball. What is the probability that it is a white ball (*W*)?

**Solution:** Note that this is a random experiment made up of two sub-experiments. The 1st is the draw of the box and the 2nd is the draw of the ball. In addition, the ball is drawn from the box, so it is attached to the box. We can therefore consider event $A$ to be the drawing of the ball, and events $B_i$ to be the drawing of one of the three boxes, each of which is completely unrelated to the other, or to its contents (*i.e.* incompatible two by two).

Therefore, we can easily apply our global probability law for event $A$ (**White** ball).

$$p(W) = \sum_{i=1}^{3} p(Bt_i) \times p_{Bt_i}(W)$$

$$= \frac{1}{3} \times \frac{5}{10} + \frac{1}{3} \times \frac{8}{10} + \frac{1}{3} \times \frac{4}{10} = \frac{17}{30}$$

This result can be obtained from the probability tree diagram. But, it should be noted that the tree will lose its efficiency when it comes to high numbers of $B_i$ or sub-experiments.

## BAYES' THEOREM

We have seen that [1, 2]: $p(A \cap B_i) = p(B_i) \times p_{B_i}(A)$

And we know that: $A \cap B_i = B_i \cap A \quad \Rightarrow \quad p(A \cap B_i) = p(B_i \cap A)$

Therefor; we can write:

$$p(B_i \cap A) = p(A) \times p_A(B_i) \tag{4.10}$$

By combining the 1st and 3rd expressions, and we divide by $p(A)$, we get:

$$p_A(B_i) = \frac{p(B_i) \times p_{B_i}(A)}{p(A)} \tag{4.11}$$

And we have already obtained (overall probability): $p(A) = \sum_{i=1}^{n} p(B_i) \times p_{B_i}(A)$

Finally, we get:

$$\boxed{p_A(B_i) = \frac{p(B_i) \times p_{B_i}(A)}{\sum_{i=1}^{n} p(B_i) \times p_{B_i}(A)}} \tag{4.12}$$

This last equation is known as "**Bayes' theorem**".

The analysis of this equation shows that it relates the probability of a path leading to an event $A$ to the sum of the probabilities of all the paths leading to the same event. Moreover, in its expression we calculate $p_A(B_i)$ and not $p_{B_i}(A)$. We moved from the **end** to the **beginning** on the probability tree diagram. This is the **objective** of *Bayes' theorem*. It is the knowledge of the ***most probable*** source for an event, made from several possible causes (outcomes).

This theorem is widely used in predicting the history of phenomena, such as the development of a virus in the body, *etc.*

**Example:** In a society, we noticed that **10%** of workers *have university degrees* and **70%** of them are in charge of *administrative tasks*. We also noticed that **30%** of workers *without university degrees* are in charge of *administrative tasks*.

-       A worker is interviewed at random. What is the probability that he is a *non-administrator* and *does not have* a university degree?

**Solution:** We summarize the example in a probability tree diagram (Fig. **4.10**). This is not mandatory.

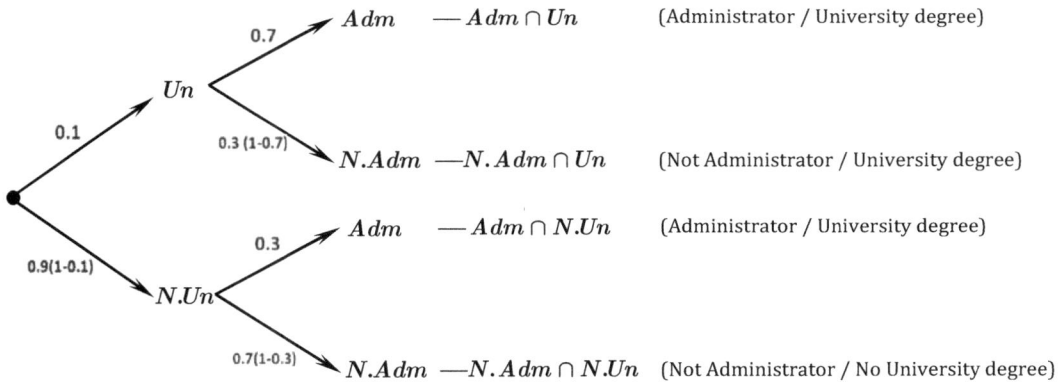

Fig. (4.10). Corresponding probability tree of the society.

From the question asked, we can see that it is about moving from the end to the beginning on the tree, by what it has been specified that it is the non-administrator first. This is therefore the case that Bayes' theorem deals with.

$$p_{N.Adm}(N.Un) = \frac{p(N.Un) \times p_{N.Un}(N.Adm)}{\sum_{i=1}^{2} p(B_i) \times p_{B_i}(N.Adm)} \qquad (B_i = Un \ \ ou \ \ N.Un)$$

We replace the values:

$$p_{N.Adm}(N.Un) = \frac{0.9 \times 0.7}{0.1 \times 0.3 + 0.9 \times 0.7} \simeq 0.954$$

This result implies that the majority of non-administrators are without a university degree. Conversely; the majority of graduates are administrators. The company has an administrative tendency. If it is a service company, that is fine. While, if it is an industrial company (of production), intervention is essential in the distribution of tasks. The reputation of the company is based on creativity, product quality, *etc.* This will lead to the importance of the quality of the workforce to be recruited, which will mainly have to be university educated. So, for such a type of company, there is a big problem in assigning tasks to the right people, hence the need for intervention to keep the company on the spot.

# CHAPTER 5

# Random Variable

**Abstract:** After the great progress in the theory of probability as a science, the passage towards more flexible and representative definitions and the possibility to be implemented numerically gave the need for a new formulation. From there, the random variable came into existence. Although random problems can be of different kinds, the random variable can be discrete or continuous, qualitative or quantitative. This chapter is dedicated to its extended definition and its exploitation in a simplified way.

**Keywords:** Bernoulli's random variable, Probability law, Quantitative and qualitative random variables, Random variable, Types of random variables.

## INTRODUCTION

In order to make probability calculations more practical, with the possibility of integrating them into calculation codes, the random variable was introduced. In a simple way, we can imagine this as the search for a representative function $f(x)$ of the outcomes of a random experiment, in which $x$ is the random variable. More details will be given in the subsequent sections.

## RANDOM VARIABLE DEFINITION

We will start with two examples in order to define what is a random variable (**RV**)?

Let's consider the random experience of throwing a two-sided token numbered 1 and 2 and a six-sided dice (from 1 to 6) [2, 3, 6].

The universe $\Omega$ of this experience is therefore couplets ($n=2\times6=12$):

$$\Omega = \{(1;1); (1;2); (1;3); (1;4); (1;5); (1;6); (2;1); (2;2); (2;3); (2;4); (2;5); (2;6)\}$$

From the universe $\Omega$, we can define infinity of random variables. Consider two illustrative examples.

**Example 01:** The random variable $x$ attached to each outcome, is equal to the *sum of the digits* of the outcome (*i.e.* the sum of the digit on the token and the one on the dice). $x$ therefore will take seven (07) possible values:

$$x = 2\;;3\;;4\;;5\;;6\;;7\;;8$$

**Example 02:** The random variable $x$ attached to each outcome is equal to the number of *even digits* in the outcome. $x$ will therefore take three (03) possible values:

$$x = 0\;;1\;;2$$

The **0** here is a value of $x$. It means that we had a result with odd digits only, for example the couplet $(1\,;5)$. It should not be forgotten therefore (*i.e.* $x=0$).

Therefore, A random variable is a function, denoted by $x$ (usually), defined on $\Omega$ and has its values in $\mathfrak{R}$ .

$$\Omega \xrightarrow{\quad x \quad} \mathfrak{R} \qquad\qquad (5.1)$$

For example 01 : $(a\,;b) \xrightarrow{\quad x \quad} a+b$

For example 02 : $(a\,;b) \xrightarrow{\quad x \quad} a \; even\, x = 1\;; b \; even \; x = 1 \;\; (or \; x = 0)$

## Sets Defined Using $x$

Let $x$ be a random variable defined on $\Omega$. The possible values of $x$ are noted $x_i$ and we can write: $\Omega \longrightarrow \{x_1\,;x_1\,; \,..... \;;x_p\}$ .

The set of antecedents of $x_i$ by $x$ is a subset of $\Omega$. This, is therefore, an event that is noted: $[x = x_i]$   *or*   $(x = x_i)$ .

With **example 01:** $[x = 4]$:The sum of the digits equal to **4** is achieved by *two couplets* which compose the event $\{(1\,;3); (2\,;2)\}$ .

We write: $[x = 4] = \{(1\,;3); (2\,;2)\}$

We can also define inequalities on random variables.

$[x < 4]$: The sum of the digits is less than **4**.

This definition requires the use of knowledge already acquired during previous courses.

We can write: $[x<4]=[x=2]\cup[x=3]$

Therefore: $[x<4]=\{(1\,;1)\}\cup\{(1\,;2);(2\,;1)\}=\{(1\,;1)\,;(1\,;2)\,;(2\,;1)\}$

## Probability Law of a Random Variable $x$

For each value $x_i$ of $x$, we have a corresponding probability. Writing the values $x_i$ affected by their corresponding probabilities $p_i$ in a table is called the "***probability law of $x$***" (Table **5.1**). This is the case with a *discrete random variable* of course (spaced $x_i$).

**Table 5.1. General form of the law of x.**

| $x$ | $x_1$ | $x_2$ | ............... | $x_p$ | - |
|---|---|---|---|---|---|
| $p[x=x_i]$ | p₁ | p₂ | ............... | p$_p$ | $\sum_{i=1}^{p} p_i = 1$ |

With **example 01**: $[x=4]=\{(1\,;3);(2\,;2)\}$

$$p[x=4]=\frac{card[x=4]}{card(\Omega)}=\frac{card\{(1\,;3);(2\,;2)\}}{card(\Omega)}=\frac{2}{12}=\frac{1}{6}$$

Following the same logic to other $x_i$ values, we arrive at the following Table (**5.2**):

**Table 5.2. Law of x of Example 1.**

| $x$ | 2 | 3 | 4 | 5 | 6 | 7 | 8 |
|---|---|---|---|---|---|---|---|
| $p[x=x_i]$ | $\frac{1}{12}$ | $\frac{1}{6}$ | $\frac{1}{6}$ | $\frac{1}{6}$ | $\frac{1}{6}$ | $\frac{1}{6}$ | $\frac{1}{12}$ |

With **example 02**: $[x=0]=\{(1\,;1);(1\,;3);(1\,;5)\}$   $\Rightarrow$   $p[x=0]=\frac{3}{12}=\frac{1}{4}$

Following the same logic to other $x_i$ values, we obtain the probability law (Table **5.3**):

**Table 5.3. Law of x of Example 2.**

| $x$ | 0 | 1 | 2 |
|---|---|---|---|
| $p[x = x_i]$ | $\dfrac{1}{4}$ | $\dfrac{1}{2}$ | $\dfrac{1}{4}$ |

## Notes:

- When you are asked to write the probability law of a random variable, you must determine the values of $x$ and their corresponding probabilities, then draw and complete the table as detailed above;
- You must check that the sum of the probabilities equals **1**, to be sure that you have swept away all the outcomes of the random experiment. Otherwise, you *forgot some*, or counted *some more than once*.

## MATHEMATICAL EXPECTATION (EXPECTED VALUE)

We define by $\overline{x}$ *(or $E(x)$)*, the average of the values taken by $x$ each assigned to its corresponding probability [2, 3, 7]. This is a weighted average and not an arithmetic average.

$$\overline{x} = \frac{p_1 \times x_1 + p_2 \times x_2 + \ldots\ldots + p_p \times x_p}{p_1 + p_2 + \ldots\ldots + p_p} \tag{5.2}$$

$$\text{As: } p_1 + p_2 + \ldots\ldots + p_p = \sum_{i=1}^{p} p_i = 1$$

We write:

$$\boxed{\overline{x} = \sum_{i=1}^{p} p_i \times x_i} \tag{5.3}$$

The calculation of mathematical expectations has several purposes. The most important is to know the average value $\overline{x}$ of the random variable, around which the other values $x_i$ are distributed. This value is an expected value if we carry out a new random experiment, governed by the weights in probability of the outcomes.

**Example 03:** In this example, we will show one of the feats of knowledge of $\overline{x}$ .

We go back to *Example* 02 (where $x$ was the number of even numbers), and we offer you a betting game with these guidelines.

• The stake is **1000 AD** (to access the game); (**AD**: Algeria Dinar)

- If the player *does not obtain an even number*, he *earns* **1500AD** (*i.e.* his **1000AD+500AD**);

- If the player *rolls a single even number*, he *loses his stake* (*i.e.* he will lose his **1000AD**);

- If the player *gets two even numbers*, he *earns* **2000AD** (*i.e.* his **1000AD+1000AD**).

o Do you think this game is favorable (to the player of course)?

**Solution:** From a primary point of view, the game seems favorable for the player, as among the three (03) possibilities, we offer *two wins*, while he *only loses* his stake for the *remaining case*. But, it should always be remembered that betting games are organized for the good of the game organizer, not the player. So, was he that stupid?

To begin the resolution, we will define a new random variable called **G**, which represents the *algebraic gain* (net gain), which is the *proposed gain minus the stake*. The definition of this new random variable is because it is the *money that matters most,* not the even numbers. It is clear that we still have the same probabilities already calculated.

The probability law for the random variable **G** is as follows (Table **5.4**):

**Table 5.4. Corresponding Law of G.**

| $x$ | 0 | 1 | 2 |
|---|---|---|---|
| $G$ | +500AD | 1000AD | +1000AD |
| $p[x = x_i]$ | $\dfrac{1}{4}$ | $\dfrac{1}{2}$ | $\dfrac{1}{4}$ |

Let's calculate $\overline{G}$ :

$$\overline{G} = \sum_{i=1}^{3} p_i \times G_i = \frac{1}{4} \times 500 \; -\frac{1}{2} \times 1000 + \frac{1}{4} \times 1000 = \boxed{-125\, AD}$$

We have found that $\overline{G}$ is negative! What does this result mean to the player?

To understand this result, it must be remembered that a betting game *based on luck*, ***must be fair*** from the start and it is luck that will pay off to the player or the organizer. So, a *fair game* must have $\overline{G}$ equal **0**.

The *negative value* indicates that the *game is not fair* from the outset, and that it is for the benefit of the organizer. For each game (throwing the token and the dice), the player risks losing 125AD. Now imagine if the player plays 100 times, or we have 100 or more players, who play -each- several times (*object of the large numbers' law, which postulates that repeating the random experiment a large number of times, will orient the results towards the value of the mathematical expectation*). Obviously, a *huge* and *easy win* will have the organizer. *This is the reason behind the rapid development and fictional profits of gambling and betting houses (unfair luck)*.

The analysis of the cause of obtaining a negative $\overline{G}$, shows that the rules of the game led to this. We propose the loss on the most probable random variable (***p=1/2***). So, the organizer is well aware of the effect of ***probability*** value, on which he has set the conditions intelligently.

For this game, it is qualified as ***unfavorable*** by a clear risk of loss. There are three cases:

○  $\overline{G} < 0$ : The game is ***unfavorable*** to the player;

○  $\overline{G} = 0$ : The game is ***fair***;

○  $\overline{G} > 0$ : The game is ***favorable*** to the player.

## VARIANCE AND STANDARD DEVIATION

In many situations (to be reviewed later), the mathematical expectation value is not enough to give a good judgment. In addition, the distribution of random variable values ($x_i$) around it, makes neglecting the density of their distributions very inappropriate [1, 3, 7]. From here, the scientists proposed two other judgment parameters (actually only one). We talk about Variance $V(x)$ and Standard deviation $\sigma(x)$.

Note that the ***choice*** of the variance formula is based on the calculation of the average of the deviations from the average (mathematical expectation) affected by their corresponding probabilities. Since the values of the random variable around the expected value can be negative or positive (*the value of their mean can therefore be zero*), the variance is calculated by *squared* deviations. Then, the root of the mean deviation gives the standard deviation which is *a measure of dispersion* around the mathematical expectation. This standard deviation is very useful in judgment as we will see later.

We give:

$$V(x) = \sum_{i=1}^{p} \left( x_i - \overline{x} \right)^2 \times p_i \qquad (5.4)$$

We can write it otherwise (obtained from the previous one[1]):

$$V(x) = \sum_{i=1}^{p} p_i \times x_i^2 - \overline{x}^2 \qquad (5.5)$$

Finally:

$$\sigma(x) = \sqrt{V(x)} \qquad (5.6)$$

For the example of the betting game, we obtain (Table **5.5**):

**Table 5.5. Calculation procedures.**

| $x$ | 0 | 1 | 2 | | |
|---|---|---|---|---|---|
| $G$ | +500AD | 1000AD | +1000AD | | |
| $p[x=x_i]$ | $\dfrac{1}{4}$ | $\dfrac{1}{2}$ | $\dfrac{1}{4}$ | | |
| $G_i \times p_i$ | +125 | -500 | +250 | $\overline{G} = -125\,AD$ | $\overline{G}^2 = 15625\,AD^2$ |
| $G_i^2 \times p_i$ | 62500 | +500000 | 250000 | $V(G) = 796875\,AD^2$ | $\sigma(G) \approx 892.68\,AD$ |

**Example 04:** Consider the two statistical series, describing the marks of two students (Mohamed and Ali), summarized in the following Table **5.6**:

$$V(x) = \sum_{i=1}^{p}\left(x_i - \overline{x}\right)^2 \times p_i = \sum_{i=1}^{p}\left(x_i^2 - 2.x_i.\overline{x} + \overline{x}^2\right) \times p_i = \sum_{i=1}^{p} x_i^2 \times p_i - \overline{x}.\underbrace{\sum_{i=1}^{p} 2.x_i \times p_i}_{\overline{x}} + \overline{x}^2$$

$$= \sum_{i=1}^{p} x_i^2 \times p_i - \overline{x}^2$$

**Table 5.6. Mark statistics of the two students.**

**Mohammed**

| Module | French | English | Math | Physics | Chemistry | Arabic literature |
|---|---|---|---|---|---|---|
| $X_M$ (Med's marks) | 12.0 | 13.0 | 14.5 | 16.0 | 12.0 | 11.5 |
| Coefficient ($n_i$) | 1 | 1 | 4 | 3 | 2 | 3 |

**Ali**

| $X_A$ (Ali's marks) | 7.0 | 5.0 | 18.5 | 17.5 | 10.5 | 10.0 |
|---|---|---|---|---|---|---|

## Notes:

- In statistics, mathematical expectation is called the mean ($\overline{X}$), because the random variables values are known at this point. However, in probability, we make calculations to predict the result (its probability of occurrence), as in the case of a game not yet played;
- In this example, the equivalent of Card($\Omega$), which is the number of possible cases is named $N$, which is the sum of the coefficients of all the modules. We can make an analogy with what has been seen in probability where: $p_i = \dfrac{Card(x_i)}{Card(\Omega)}$, while here: $p_i = \dfrac{Coef(n_i)}{N}$.

Let's now calculate the averages for the two students:

$$\overline{X}_M = \frac{1}{N}\sum_{i=1}^{6} n_i \times x_{Mi} = \frac{1}{14}\left(1\times12.0 + 1\times13.0 + 1\times14.5 + 1\times16.0 + 1\times12.0 + 1\times11.5\right) \approx \boxed{13.54}$$

$$\bar{X}_A = \frac{1}{N}\sum_{i=1}^{6} n_i \times x_{Ai} = \frac{1}{14}\left(1\times7.0+1\times5.0+1\times18.5+1\times17.5+1\times10.5+1\times10.0\right) \simeq \boxed{13.54}$$

We can see that the averages of both students are *equal* (*stroke of luck!*). From a point of view purely based on their averages, we can say that the two students are equivalent.

Now, we will calculate their variances, and then their standard deviations (*we keep the same names in statistics*):

$$V_M(X) = \frac{1}{N}\sum_{i=1}^{6} n_i \times x_{Mi}^2 - \bar{X}_M^2 = \frac{1}{14}\left(1\times12.0^2+1\times13.0^2+1\times14.5^2+1\times16.0^2+1\times12.0^2+1\times11.5^2\right)-13.54 \simeq 2.98$$

$$V_A(X) = \frac{1}{N}\sum_{i=1}^{6} n_i \times x_{Ai}^2 - \bar{X}_A^2 = \frac{1}{14}\left(1\times7.0^2+1\times5.0^2+1\times18.5^2+1\times17.5^2+1\times10.5^2+1\times10.0^2\right)-13.54 \simeq 22.66$$

$$\sigma_M(X) = \sqrt{V_M(X)} \simeq \boxed{1.72}$$

$$\sigma_A(X) = \sqrt{V_A(X)} \simeq \boxed{4.76}$$

We see that the variances and precisely the standard deviations of the two students are different, despite their averages being equal.

The value of the standard deviation is very useful in statistics. It measures the degree of dispersion around the average. For the present example, we see that Mohammed's grades are more localized close to his average, which means that overall he is a good student who performs well in all modules. On the other hand, for *Ali*, the marks are very dispersed with respect to his average, which means that he is a strong student only in a few modules (with high coefficients). So, his career will depend a lot on his report card (he has a very technical tendency), while for *Mohammed* the choices are more extensive.

The interest from this example will spread to other areas, namely: population growth, temperature variation during a period, rainfall, and so on.

# QUANTITATIVE AND QUALITATIVE RANDOM VARIABLES

For a random variable $x$ of any random experiment, we distinguish those for which we are interested in values (quantities). We thus speak of a **quantitative random variable** [2, 7].

## Examples

- We roll a dice, and we look at the apparent number on the upper facet. If we denote the result obtained by $x$, the values of $x$ are: 1; 2; 3; 4; 5; 6;
- If we roll two dice successively, and we denote by $x$, the sum (or the sum squared, *etc.*) of the numbers obtained. $x$ will take the values 2; 3; 4; 5; 6; 7; 8; 9; 10; 11; 12;
- The experiments (statistics) concerning the measurements (scores, temperature, rainfall,…) are all based on random variables described by values.

When we are only interested in a quality (a characteristic), designated by a random variable $x$, we say that it is **qualitative**.

## Examples

- We roll a dice, and we are interested in the even number ($x=1$ for **even** and **0** for **odd**);
- We roll two dice, and we denote by $x$ the number of even digits ($x=0$ or **1** or **2**)
- We draw a ball from an urn containing 22 balls: red, green, and white numbered from 1 to 22. When we are interested in color only, it is a qualitative variable.

## Bernoulli's Random Variable

Suppose we have an urn containing **black** and **white** balls. We denote by $N_B$ the number of *black balls*, and by $N_W$ the number of *white balls*.

The total number of balls in the urn is: $N = N_B + N_W$

Suppose a ball has been drawn from the urn. Clearly, it will be *black* or *white*. So, we can assume that the *color* is a random variable (qualitative), which can take the color *black* or the color *white*.

So, we have:

$$\Omega = \{B \; ; W\} \tag{5.7}$$

$$P(x=W)=\frac{N_W}{N_B+N_W}=p \qquad ; \qquad P(x=B)=\frac{N_B}{N_B+N_W}=q \qquad (5.8)$$

*p* is chosen for the event we are interested in, by definition.

It is obvious that: $P(x=B)=1-\dfrac{N_W}{N_B+N_W}=1-p$

**Note:** In general, *Bernoulli*'s random variable is concerned with ***opposing*** (contrary) events. For example: *Black* and *white*, *success* and *defeat*, *male* and *female*,… *etc.*

For this, we assign for these two events, two values of *x*. the values **0** for *Black* and **1** for *White* (same for the other examples cited above).

So, for our urn, we can write the law of probabilities of *x* as Table (**5.7**):

**Table 5.7. Probability law using Bernoulli's random variable definition.**

| x (ball color) | W | B |
|:---:|:---:|:---:|
|  | 1 | 0 |
| P(X=x$_i$) | p | q |

## Types of Random Variables

There are *two main* types of random variables:

- Discrete random variables (discontinued or separated);

- Continuous random variables (attached).

### *Discrete Random Variable*

It is of *two kinds*:

### Discrete Random Variable with Finite Universe

It is the random variable that can take a finite number of integer values (non-rational), in its universe of variation ($\Omega=\{x_1; x_2; …; x_n\}$).

The examples for this case are various:

- Throwing a dice, where $\Omega = \{1; 2; 3; 4; 5; 6\}$

- The number of visitors to an art museum, where $\Omega = \{0; 1; 2; ...; n\}$

## Discrete Random Variable with Countable Infinite Universe

It is the random variable which can take an infinite number of integer values, *i.e.* that the set $x(\Omega)$, is countable with a lower element. The best way to understand this type of variable is the following example.

## Example 05

Suppose an urn containing white balls (**W**) with percentage (or probability) **p**, and black balls (**B**) with percentage **q** (**q=1-p**). We want to draw a **white ball**. The experiment can therefore be completed in the *first draw* if the drawn ball is *white*. But, if the *drawn ball* is *black*, another draw is made after returning the black ball to the urn. *We repeat* the drawing operation until we obtain our desired *white ball*. So, the *experience ends* once the *desired ball is drawn*. The number of draw operations required -which can range from **1** to **n**- is just a *random variable* that does not depend on the number of balls of course.

We calculate now the series of probabilities of the values of *x*:

$$
\begin{aligned}
p_1 &= P(x=1) = p \\
p_2 &= P(x=2) = q.p \\
p_3 &= P(x=3) = q.q.p = q^2.p \\
\vdots \quad &\quad \vdots \quad\quad\quad \vdots \\
p_n &= P(x=n) = q^{n-2}.q.p = q^{n-1}.p
\end{aligned}
\tag{5.9}
$$

If, for example, there are *equal numbers* of white and black balls in the urn, their probabilities for a single draw are:

$$
p = \frac{N_W}{N_W + N_B} = q = \frac{N_B}{N_W + N_B} = \frac{1}{2} \quad (N_W = N_B)
\tag{5.10}
$$

So, the series of probabilities of *x* will be:

$$p_1 = \frac{1}{2}$$

$$p_2 = \frac{1}{2} \cdot \frac{1}{2} = \frac{1}{2^2}$$

$$p_3 = \frac{1}{2^2} \cdot \frac{1}{2} = \frac{1}{2^3}$$

$$\vdots \qquad \vdots \qquad \vdots$$

$$p_n = \frac{1}{2^{n-1}} \cdot \frac{1}{2} = \frac{1}{2^n}$$

**(5.11)**

Hence, the probability of having a *white ball* decreases with **n** increase, but always remains greater than **0**.

It remains to be verified that the sum of the probabilities of *x=1* up to *x=n* is indeed **1**.

$$\sum_{i=1}^{n} p_i = p + p.q + p.q^2 + ... + p.q^{n-1} \quad (\textbf{\textit{n}} \text{ can go up to } +\infty)$$

**(5.12)**

We bring out **p** as a factor:

$$\sum_{i=1}^{n} p_i = p\left(1 + q + q^2 + ... + q^{n-1}\right)$$

**(5.13)**

The expression in parenthesis is nothing other than a *geometric sequence* of reason **q**. So, the sum of its terms is done according to the known rule:

$$1 + q + q^2 + ... + q^{n-1} = 1 \times \frac{1-q^n}{1-q} \qquad \left(q^n \xrightarrow[n\to\infty]{} 0\right)$$

**(5.14)**

Or:

$$1 + q + q^2 + ... + q^{n-1} \simeq \frac{1}{1-q}$$

**(5.15)**

And, we know that:  **p=1-q**

So we finally have:

$$\sum_{i=1}^{n} p_i = p\frac{1}{1-q} = \frac{p}{p} = 1$$

**(5.16)**

The law of probabilities of $x$ (table) is written in this case as follows (Table **5.8**):

**Table 5.8. Law of the discrete RV with countable infinite universe.**

| $x$ | 1 | 2 | 3 | ... | $n$-1 | $n$ |
|---|---|---|---|---|---|---|
| $P(x=x_i)$ | $p$ | $q.p$ | $q^2.p$ | ... | $q^{n-2}.p$ | $q^{n-1}.p$ |

## *Continuous Random Variable*

We say that a *random variable* is **continuous**, when it takes an infinite number of possible values over an *interval* of values, and represented by the *area under a curve*.

## **Example 06**

In the experiment to estimate the rainfall (measured in *mm*. 1.0 *mm* = 1.0 *liter/m²*) in the coming days, we can make lower and upper limits [**0mm** ; **50mm**]. The first limit (**0mm**) means it doesn't rain at all, while the second (**50mm**) is an excess limit (according to recorded statistics). We know that the falling quantity will be between the two limits. In addition, if it is measured, the next day, the weather forecast will say that it is, for example, **7.0 *mm***. However, in reality the quantity is not exactly **7.0 *mm***, it is a *little more* or a *little less* (a *rain molecule* has its *own volume*). With more precision, we can say without being wrong, that the quantity is, for example **7.00000125.... *mm***. For such a case, we can say that the *amount of falling rain* is a *continuous random variable*, since it takes any value between two limits without being able to quantify it exactly.

If now the rainfall statistics are made on every day of the year for a long period, we will have a representative curve (Fig. **5.1**), for which, the random variable $x$ is continuous.

This example can be made on the change of temperature, humidity, ...*etc*. It can also be done to measure the height (size) of a country's population, their lifespan, their weight, ... *etc*.

Therefore, and as a general rule, the continuous random variable is a random variable, whose value cannot be determined with exactitude, and it is between two values.

**Note:** If, for the previous example, we say that we are interested in rainfall with an accuracy of 1%, *i.e.* the third digit is round off after the decimal point. In this case, we switch to a *discontinuous* (*discrete*) *random variable*.

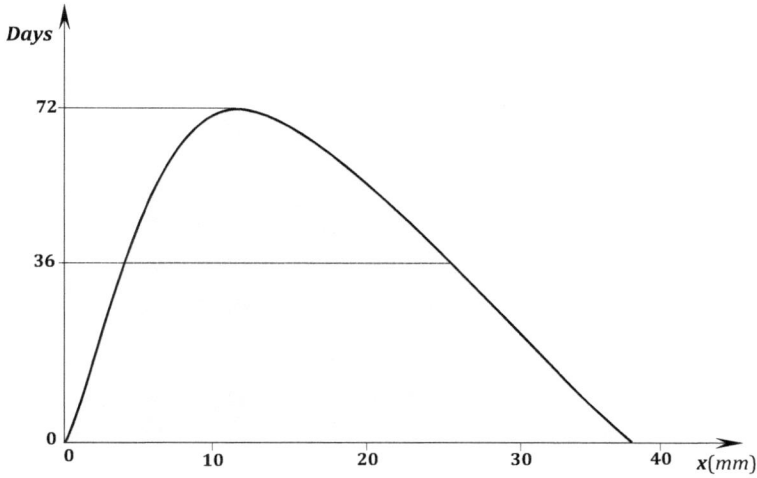

**Fig. (5.1).** Average annual rainfall (40 years) for Algiers region (Algeria's capital).

# CHAPTER 6

# Distribution Laws

**Abstract:** In this last chapter of the probability part, the laws of probability are presented. These laws are very important even in statistics. They govern difficult phenomena by simple mathematical formulas, obtained after a few developments. The mastery of these laws will allow the classification of the phenomena by the amounts of the probabilities and the number of elements in the samples (or the population).

**Keywords:** Binomial Law, Central limit theorem, Distribution function, Normal law, Poisson's Law, Uniform law.

## INTRODUCTION

Like any field, having formulas that are simple to use, instead of going through calculation codes - sometimes essential - is well desired. This implies the formulation of laws of variation of the random variable *x*, often called ***distribution laws***. They are generally valid for certain ranges of variation of *x*, or for certain characteristics of the random experiment itself.

## DISTRIBUTION FUNCTION (CUMULATIVE DISTRIBUTION FUNCTION (CDF))

The distribution function of a random variable *x* (or precisely its law), is the function $F_x$ from $\mathbb{R}$ to $[0,1]$, which associates with any value of *x*, the probability of obtaining a value less than or equal to the value of *x* considered [1, 4, 6].

- For a discrete random variable:

$$F_x = p\left(x \le x_i\right) = \sum_1^i p_i \tag{6.1}$$

We can notice that $F_x$ represents the cumulative probability of *x*.

- For a continuous random variable:

$$F_x(x) = \int_{-\infty}^x f_x(y).dy \tag{6.2}$$

Where $f_x$ (or $p(x)$)represents the density function of $x$, which describes its law.

## Discrete Random Variable

For the case of a *discrete random variable*, the distribution function leads to a *staircase-shaped* evolution:

Consider the following distribution law of $x$ Table **6.1**.

**Table 6.1. Distribution function calculation.**

| $x$ | -3 | 0 | 3 | 6 |
|---|---|---|---|---|
| $P(x=x_i)$ | 1/8 | 3/8 | 3/8 | 1/8 |
| $F_{xi}$ | 1/8 | 4/8 | 7/8 | 1 |
|  | $\sum_1^1 p_i = 1/8$ | $\sum_1^2 p_i = p_1 + p_2$ $= 1/8 + 3/8$ $= 4/8$ | $\sum_1^3 p_i$ | $\sum_1^4 p_i$ |

## *Graphical Representation*

The *probability* is represented graphically by *bars*, while the *distribution function* is represented by *horizontal segments* (Fig. **6.1**). Note that for $F_x$, we have the value **0** and the value **1** *outside* the domain of the experiment (before and after). This can be explained as follows: before $x_1$ ($x < x_1$) we have a *zero probability* (impossible event or result) while after $x_4$ ($x > x_4$) all possible results are achieved (certain event or result).

## Binomial Law

It is based on the Bernoulli random variable (*p* for success or achievement and *q* for failure or non-achievement). Its objective is to determine the probability of a precise result for an experiment of *n sub-experiments*.

# Example 01

A *fake coin* is thrown with a probability of *tail* (*T*) realization equal to **0.4** (here *tail* corresponds to *p* or desired result). Therefore, the probability of making *head* (*H*) is equal to **0.6** (*q* for head). We throw the coin five times in a row.

-   What is the probability of having two heads followed by three tails (two-times head followed by three-times tails exactly)?

-   What is the probability of having two heads and three tails?

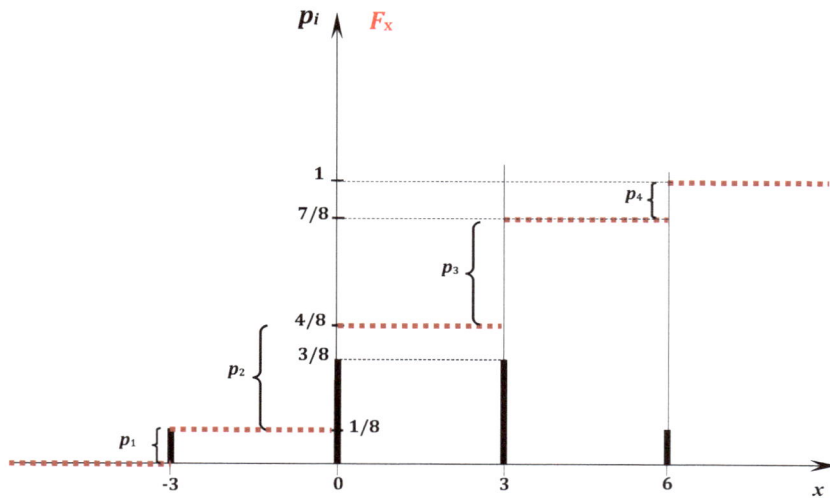

**Fig. (6.1).** Graphical representation of probabilities and distribution function of *x*.

# Solution

•   For *the first question* and because the result of a throw is independent of the one before and after, we can easily calculate the probability by multiplying the probabilities for each throw:

$$p(HHTTT) = p(H) \times p(H) \times p(T) \times p(T) \times p(T) = 0.6 \times 0.6 \times 0.4 \times 0.4 \times 0.4 = 0.6^2 \times 0.4^3$$

If we are most interested in the *tail* event, then we have three tails. We can say we have *k* tails among *n* launches. Furthermore; if the event is not *tail,* it is necessarily *head*. So, for *n* launches, if we have *k* tails, we necessarily have (*n-k*) heads.

We can therefore write that the preceding probability is equal to:

$$p(HHTTT) = p(x = k = 3) = p^k \times q^{(n-k)}$$

So, for **n**=5 and **k**=3 with **p**=0.4 and **q**=1-**p**=0.6.

$$p(HHTTT) = p(x = k = 3) = 0.4^3 \times 0.6^{(5-3)} = 0.4^3 \times 0.6^2 = 0.6^2 \times 0.4^3$$

- For *the second question*, we are not interested in the order but only in outgoing events. So, the problem goes to making permutations between the events (**H** and **T**). Because we have repetitions, we must divide by the permutations of repeated cases (review the combinatorial analysis chapter).

So :
$$N = \frac{5!}{2! \times 3!} \left( \frac{Perm\ without\ rep}{Perm\ rep\ of\ H \times Perm\ rep\ of\ T} \right)$$

We can write this last expression as follows:

$$N = \frac{5!}{2! \times 3!} = \frac{n!}{(n-k)! \times k!} = \frac{n!}{k! \times (n-k)!} = C_n^k$$

So, the sought probability is:

$$p(2H\ and\ 3T) = C_5^2 \times p^3 \times q^{5-2}$$

This last writing can be generalized according to the expression:

$$p((n-k)H\ and\ kT) = C_n^k \times p^k \times q^{n-k} \qquad \textbf{(6.3)}$$

The number of trials **n** and the probability of success **p**, are called: ***Parameters of the binomial distribution***.

Therefore, we say that a *random variable* follows a *binomial distribution* of parameters **n** and **p** and we write:

$$\boxed{x \to \beta(n, p)} \qquad \textbf{(6.4)}$$

With: The probability of **k** successes:

$$p(x = k) = C_n^k \times p^k \times q^{n-k}$$          **(6.5)**

## Example 02

A high school has **50 teachers**. The probability that a teacher *falls sick* in any day is estimated to be **0.01**.

For any day:

1.What is the probability that *no teacher* is sick?;

2.What is the probability that **5** *teachers* are sick? ;

3.What is the probability that *at least* **1** *teacher* is sick? ;

4.What is the probability that *at most* **3** *teachers* are sick?

## Solution

We can see from this example that it can be modeled by Bernoulli's random variable where: *S* represents *sick* (for success: an event that interests us) and *F* represents *non-sick* (for failure). Moreover, *p(S)*=0.01 and *p(F)*=0.99.

It is clear that the teachers can be considered as *independent events*.

So, the values of *x* are: $x = \{0;1;2; .......;49;50\}$

*i.e.* **no teacher** is sick; **only 1** is sick; **02** are sick; .... **49** are sick; **all of them** are sick on the same day.

The law parameters *n* and *p*, are *n*=50 and *p*=0.01. We write: $x \rightarrow \beta(50,0.01)$

The answers to the different situations are:

1.We apply the Binomial law with *k*=0 for this question:

$$p(x = k = 0) = C_n^k \times p^k \times q^{n-k} = C_{50}^0 \times p^0 \times q^{50-0} = \frac{50!}{0! \times 50!} \times 0.01^0 \times 0.99^{50} = 1 \times 1 \times 0.99^{50} \approx 0.605$$

2.For **5 teachers (*k*=5)**:    $p(X = 5) = C_{50}^5 \times p^5 \times q^{45} = \frac{50!}{5! \times 45!} \times 0.01^5 \times 0.99^{45}$

**3.** For **at least 1 teacher:** *i.e.* **1** or **2** or ...... or all the **50**.

It is preferable to work with the complementary: $p(X \geq 1) = 1 - p(X = 0) \approx 1 - 0.605$

**4.** For **at most 3 teachers:**

$$X \leq 3 \Rightarrow p(X \leq 3) = p(X = 0) + p(X = 1) + p(X = 2) + p(X = 3)$$
$$= C_{50}^0 \times p^0 \times q^{50} + C_{50}^1 \times p^1 \times q^{49} + C_{50}^2 \times p^2 \times q^{48} + C_{50}^3 \times p^3 \times q^{47}$$

**Properties:**

$$\boxed{\begin{aligned} &E(X \to \beta(n, p)) = n \times p \\ &V(X \to \beta(n, p)) = n \times p \times (1 - p) \\ &\sigma(X \to \beta(n, p)) = \sqrt{n \times p \times (1 - p)} \end{aligned}}$$ **(6.6)**

For the precedent example:

$$E(X \to \beta(50, 0.01)) = 50 \times 0.01 = 0.5$$

$$V(X \to \beta(n, p)) = 50 \times 0.01 \times (1 - 0.01) = 0.495$$

$$\sigma(X \to \beta(n, p)) = \sqrt{0.495} \approx 0.703$$

**Important note:** The adjective "binomial", comes from the fact that when we *sum all these probabilities*, we find the development of Newton's **binomial**:

$$\boxed{\sum_{k=0}^n C_n^k \times p^k \times q^{n-k} = (p + q)^n = 1}$$ **(6.7)**

We can check for **n**=1 and **n**=2.

### *Graphical Representation of the Binomial Distribution*

It is usually presented as a bars graph. Since the law depends on **n** and **p**, we will have various graphical representations if we vary **n** and/or **p**, as in the following cases (Fig. **6.2**).

Several remarks can be made about these diagrams (sub-figures):

- The shape of the distribution is *symmetric* if **p**=0.5, whatever the value of **n**;

- It is *asymmetric* in the case where $p{\neq}0.5$. If $p{<}0.5$, the probabilities are **higher** on the **left side** of the distribution than on the right side (*positive asymmetry*). If $p{>}0.5$, it is *the reverse* (*negative asymmetry*);

- The distribution *tends to become symmetrical* (and nearly *continuous*) when **n** becomes large. Moreover, if **p** is **far from 0** or **1**, it will approach the distribution of the *normal law*, which we will see later.

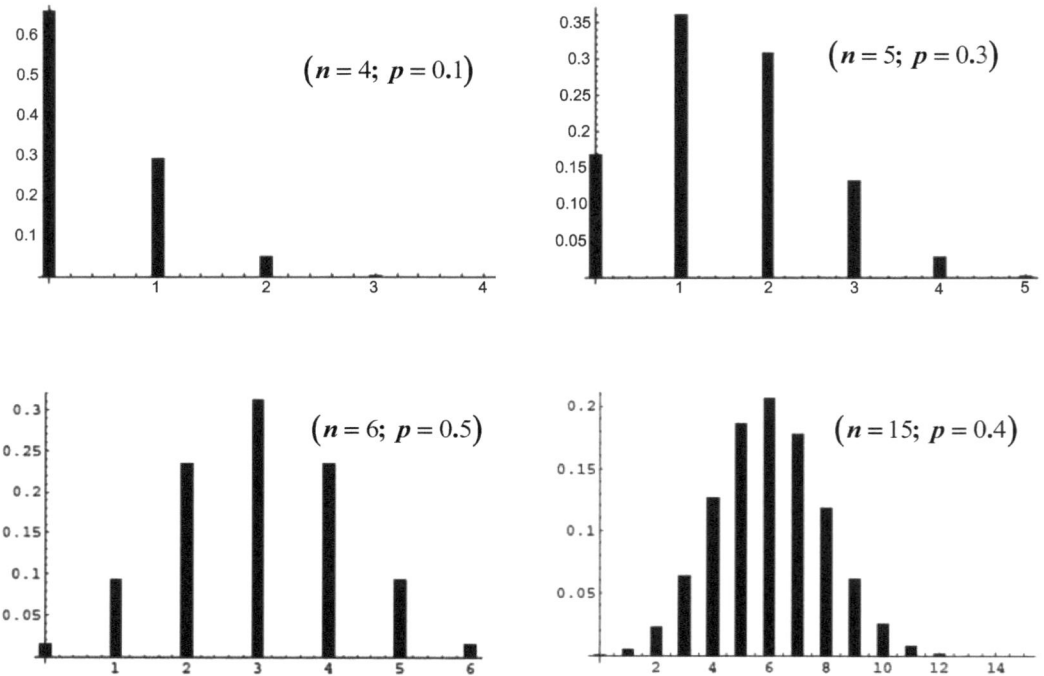

Fig. **(6.2).** Some distributions of the binomial law.

## Other laws of Distribution

In what follows, we will list without too much detail other laws of probability. Some laws are hardly used, while others are *often encountered* (Poisson law and normal law). The objective is to allow the reader to know them even in summary. It should be noted that a good understanding of the Binomial law helps very well to understand them.

## *Uniform Law*

As its name suggests, it is characterized by equal probability for all the values taken by the random variable $x$. We are in the case of an equiprobability, where $p$ $(x=x_i)$ $=1/n$, with $n$ the number of $x_i$.

For this law:

$$E(X) = \frac{n+1}{2} \quad ; \quad V(X) = \frac{n^2-1}{12} \tag{6.8}$$

**Example 03:** The distribution (or the law of the RV) of the numbers, obtained from the roll of a non-rigged dice, follows a uniform law as shown in the table below.

| $x$ | 1 | 2 | 3 | 4 | 5 | 6 |
|---|---|---|---|---|---|---|
| $p(x=x_i)$ | 1/6 | 1/6 | 1/6 | 1/6 | 1/6 | 1/6 |

$$E(X) = \frac{1}{6}\sum_{i=1}^{6} x_i = \frac{21}{6} = \frac{7}{2} = \frac{n+1}{2}$$

$$V(X) = \frac{1}{6}\sum_{i=1}^{6} x_i^2 - E(X)^2 = \frac{91}{6} - \frac{49}{4} = \frac{35}{12} = \frac{n^2-1}{12}$$

## *Poisson's Law* (*proposed by S.D. Poisson around 1837*)

It is often applied to cases of accidental phenomena, where the probability $p$ (Bernoulli's second parameter) is very low (<0.05). It can be seen as a *limit* of *Binomial law* in some cases.

When $n$ becomes large, calculating the probabilities of a Binomial distribution becomes tedious (boring). We therefore propose the following approximation:

$$if : n \to \infty \quad \Rightarrow \quad p \to 0$$
$$x \to \beta(n,p) \longrightarrow x \to \wp(\lambda) \qquad \left|\lambda = n.p\right. \tag{6.9}$$

**Note:** This approximation is very fair for $n \geq 50$ and $\lambda \leq 50$

The *Poisson' distribution* is given by the expression:

$$x \to \wp(\lambda) \quad \Rightarrow \quad p(x = k) = e^{-\lambda} \frac{\lambda^k}{k!} \tag{6.10}$$

For this law:

$$E(x) = \lambda \qquad ; \qquad V(x) = \lambda \tag{6.11}$$

## Example 04

A population contains -on average- one person *measuring more than* **1.90m** among **80 people**. *Out of* 100 *people*, calculate the probability that there is *at least* one person over **1.90m** tall.

## Solution

The probability that *one* among 80 *people* is taller than 1.90m is very low (**p**=1/80). We can therefore use Poisson's law with the parameter $\lambda = n.p$ .

- For *at-least* one person measuring more than 1.90m out of 100 people, the use of the complementary is faster. Therefore:

$$p(x \geq 1) = 1 - p(x = 0) = 1 - e^{-100/80} \frac{\left(\frac{100}{80}\right)^0}{0!} = 1 - e^{-5/4} \simeq 0.7135$$

*Normal law* (*also known as* ***Gauss's*** *or* ***Gauss-Laplace law***)

It is the best known, if not the *most used in the field of statistics*. It is applied to all real numbers, for $x \in \,]-\infty, +\infty[$ .

We say that a random variable **x**, follows a normal distribution (Fig. **6.3**) of *parameters* $\mu$ (expectation or mean) and $\sigma$ (standard deviation), and is described by:

$$x \to N(\mu, \sigma) \Rightarrow p(x) = \frac{1}{\sigma\sqrt{2\pi}} e^{-\frac{1}{2}\left(\frac{x-\mu}{\sigma}\right)^2} \quad -\infty < x < +\infty \mid p(x) \; or \; f(x) \tag{6.12}$$

**Notes**

- When $\mu = 0$ and $\sigma = 1$, it is *often called* **standard normal law**, or *reduced* **centered normal law** ( $p(x) = e^{-x^2/2}/\sqrt{2\pi}$ ). It is *very useful* for the *computation* of the standard deviation by exploiting the **Z-table** ( $z = (x - \mu)/\sigma$: a change of variable which leads any normal distribution to be a standard normal distribution). This will be detailed in the statistics part of this book;

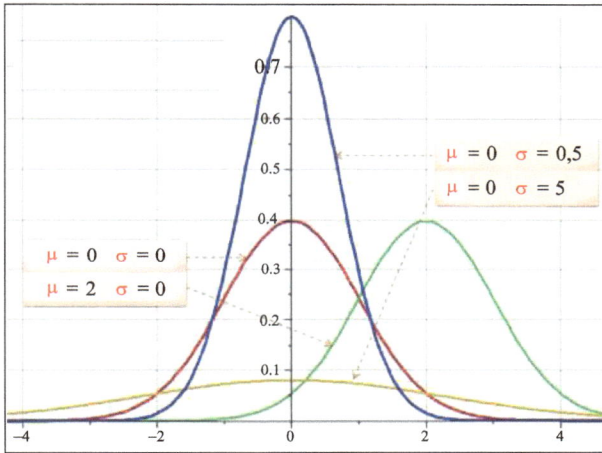

**Fig. (6.3).** Effect of *μ* and *σ* on the shape of the normal distribution.

## CENTRAL LIMIT THEOREM

This theorem is *one of the pillars* of probability theory. It *provides a simple method* for calculating probabilities related to a set of random variables. It also explains the remarkable fact that many natural phenomena follow a distribution having the shape of a *bell curve*, that is, a *normal distribution* (Fig. **6.4**) [5, 6]. Thanks to the rigorous demonstrations of the Russian mathematician *Alexander Liapunov* (1902), that we know that *all distributions converge* to *normal distribution* when *n* becomes large. We often require *n>30*, but the best approximation requires larger value, especially for distributions with other influential parameters (Bernoulli's *p* for example).

So, in summary, if we have to study a population, it suffices to take a sample of elements of the population greater than 30 (precautionary measure), and study it as a *normal distribution* (widely used and mastered), without needing to know the real distribution of the population.

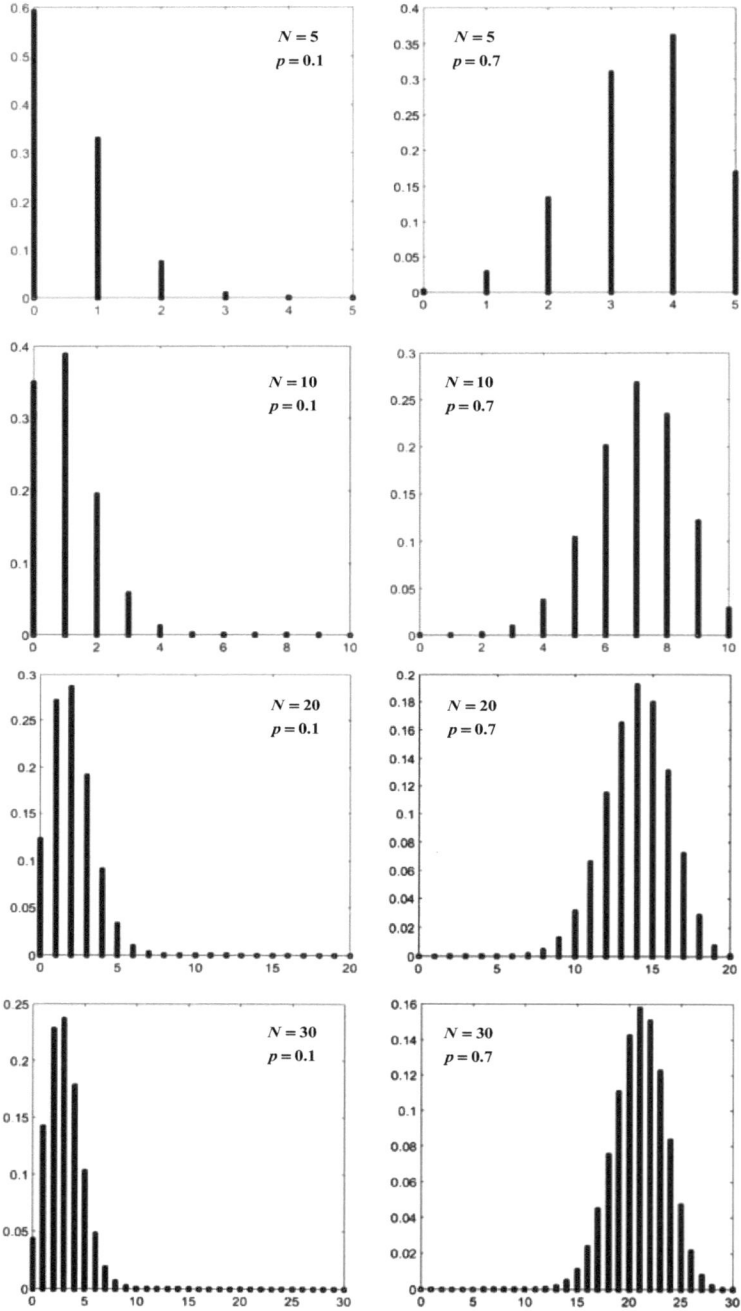

**Fig. (6.4).** Binomial-Law tendency towards Normal-Law with the increase of sample size (*N*).

In case where our population is very large, and we want to make a better evaluation, we will make several samples, each containing more than **30** elements. In this case, each sample will have its own mean and standard deviation. The mean of the means and its standard deviation, will converge to those of the normal distribution with the specifications listed. This theorem will be extensively illustrated in the *Statistics part*.

# PART 2: STATISTICS

# CHAPTER 7

# Definitions and Calculations in Statistics

**Abstract:** The Statistical part was included in only one chapter given the ample details provided for the probability part. For this, the necessary terminology in statistics is presented well as it should be. Thereafter, the exploitation of the central tendency to determine the confidence interval and the risk of error according to different situations are presented in detail and with illustrative examples.

**Keywords:** Confidence interval, Central tendency and indicators, Relative standard deviation, Risk of error, Statistical series, Vocabularies.

## INTRODUCTION

Statistics is the study of a phenomenon by collecting data, processing it, analyzing it, interpreting the results and presenting them, in order to make the data understandable by everyone [8-10]. It is both, a science and a method or a set of techniques.

Sometimes:

- The **Statistics**: With a *capital* **S** to describe the science;
- The **statistic**: With *lowercase* **s** for a statistic.

It is applied (for applied statistics) to many areas, such as:

- **Geophysics:** Weather forecasts, climatology, pollution,… *etc*;
- **Demography:** Census of a population, which gives a photograph of it, for the purpose of representative surveys sampled when necessary;
- **Physics:** By making statistics on samples that provide a clear idea of the overall behavior of physical phenomena or systems;
- …... others.

We speak of *descriptive statistics* (our case), when we are interested in collecting and processing data for understanding; While we talk about *mathematical statistics* when we are looking to find estimators for future prevention.

A statistical series can be with a ***single variable*** or with ***several variables***.

**Notice:** We can see the part of *statistics* as that of *probability*, because we have the same logic and the same laws. In probability, we work on non-existent phenomena, of which we want to determine the outcome according to the maximum probability of realization or non-outcome in the opposite case. In other words, we want to control chance. In Statistics, we work on facts, whose values we know and we try to take where there is a lack of information by probabilistic tendency (we have the same distribution laws in both parts). In probability, the achievement of an outcome is a percentage between the partial achievement and the total one (Number of favorable cases/Number of possible cases). In statistics, it is the -classical- percentage between elements (Number of elements in a category /Total number of elements).

## SOME VOCABULARIES

### Population

The set of individuals on which our statistics are based;

### Individual

An element of the population;

### Sample

Part of the population. It must be representative and sufficiently numerous. It is used when the number of the population is very large;

**Example 01:** If, for example, a statistical study is carried out on an automobile manufacturing company:

-   *The population*: Automobiles;
-   *The sample*: 100 automobiles for example,`of which all models are included;
-   *The individual*: One car.

### Modality

This is the characteristic studied for the individual. It can be quantitative (defined as measurable) or qualitative (not measurable).

We can summarize that each statistical variable (also called random variable) follows the diagram below (Fig. **7.1**):

## A STATISTICAL SERIES (WITH ONE VARIABLE) $\{(x_i, n_i)\}$

- $x_i$ : The value of the statistical variable [8;11];
- $n_i$ : The corresponding workforce (number of individuals for the same $x_i$ )

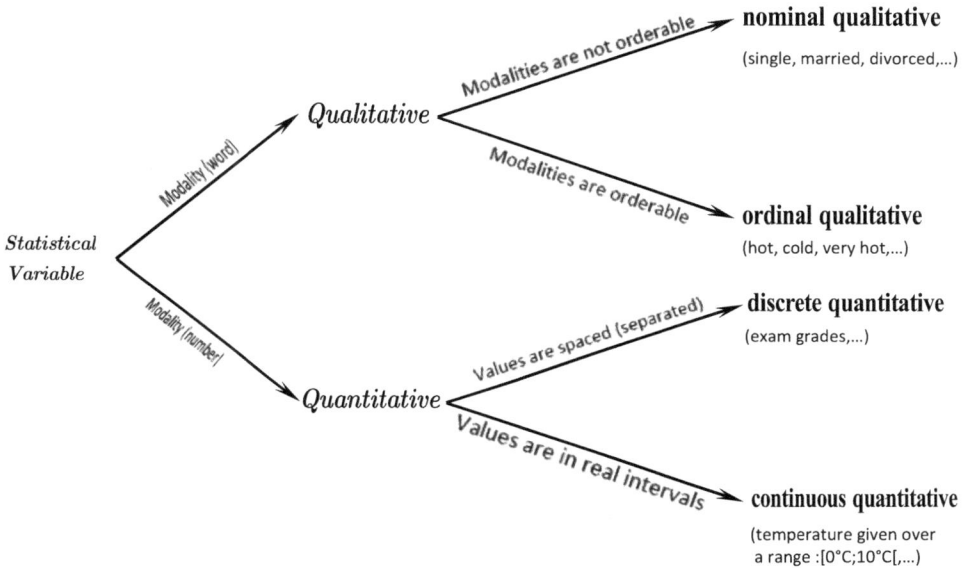

**Fig. (7.1).** Modality of the statistical random variable.

**Example 02:** We did a statistical analysis of **30** *students* about how *many siblings* (brothers and sisters) each of them has. We obtained (Table **7.1**):

**Table 7.1. Statistical series of the 30 students.**

| $x_i$ (N$^{br}$ of siblings) | 0 | 1 | 2 | 3 | 4 | 5 | |
|---|---|---|---|---|---|---|---|
| $n_i$ (N$^{br}$ of students) | 4 | 5 | 8 | 4 | 7 | 2 | $\sum\limits_{i=1}^{6} n_i = 30$ |

So we have 04 persons (students) who have no brothers or sisters at home. 02 only have five siblings.

$$N = \sum_{i=1}^{n} n_i \qquad (7.1)$$

Named the ***total workforce*** (in the example it equals 30)

➢ In statistics, we prefer to name $n_i$ the ***absolute frequency***.

➢ When we look for the percentage (the influence) of each $n_i$, we divide by $N$ (the total number) and we talk about ***relative frequency (Fr)***.

➢ We often calculate the ***increasing (ascending) cumulative*** and ***descending (descending) frequencies (ICF and DCF)***.

➢ The ***increasing (ascending) cumulative*** and ***decreasing (descending)*** <u>*relative frequencies (IC-Fr and DC-Fr)*</u> are also often calculated (Table **7.2**).

**Table 7.2. Statistical series and extended calculations.**

| $x_i$ | 0 | 1 | 2 | 3 | 4 | 5 | - |
|---|---|---|---|---|---|---|---|
| $n_i$ | 4 | 5 | 8 | 4 | 7 | 2 | $\sum_{i=1}^{6} n_i = 30$ |
| ICF (or ICW) | 4 | 9 | 17 | 21 | 28 | 30 | - |
| DCF (or DCW) | 30 | 26 | 21 | 13 | 9 | 2 | - |
| Fr | 4/30 | 5/30 | 8/30 | 4/30 | 7/30 | 2/30 | - |
| IC-Fr | 4/30 | 9/30 | 17/30 | 21/30 | 28/30 | 1 | - |
| DC-Fr | 1 | 26/30 | 21/30 | 13/30 | 9/30 | 2/30 | - |

In the previous example, the statistical variable is ***discrete*** (identical to that seen in probabilities).

For a ***continuous random variable***, we give the following example (Table **7.3**):

**Table 7.3. Statistical series for continuous RV.**

| $x_i$ (named: class $i$) | [10-20] | [20-30] | [30-40] | [40-50] | |
|---|---|---|---|---|---|
| $n_i$ (frequency or workforce) | 4 | 5 | 8 | 4 | $N = 20$ |

The same procedures are followed to calculate: ***IC-F***, ***DC-F***, ***Fr***, ***IC-Fr*** and ***DC-Fr***.

## Graphical Representation

For a *discrete statistical variable* (Fig. **7.2**), we will have ***bars***. For a *continuous variable,* we will have ***rectangles*** or ***a curve***.

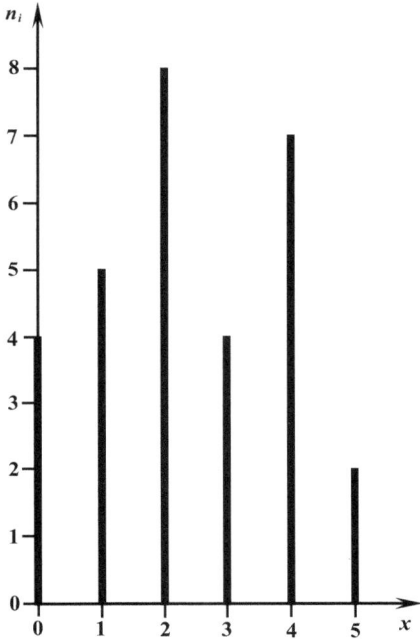

Bars' Diagram (Table 1)          Histograms' Diagram (Table 3)

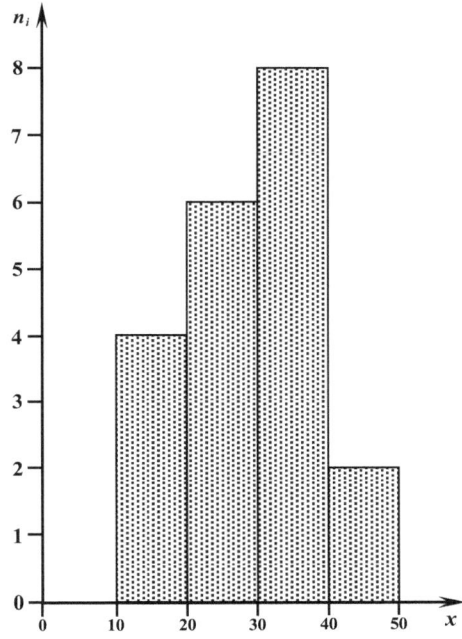

**Fig. (7.2). Diagram for discrete (left) and continuous (right) RV.**

We can use a ***sectorial (Pie chart) representation*** (Fig. **7.3**) for a *discrete statistical variable*, just as we can use **polygons** for the *discrete* or *continuous variables*. For the second representation, we place the ***nodes*** on the *vertices* or in the *middle of each rectangle* (Fig. **7.4**).

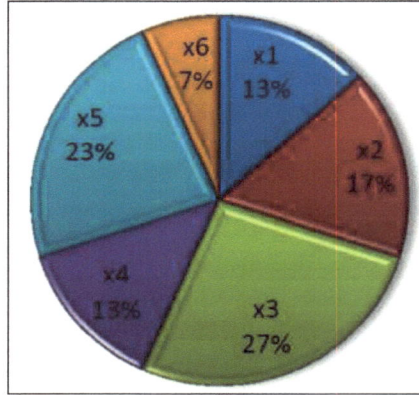

**Fig. (7.3).** Sectorial diagram (or camembert) (Table **7.1** Data).

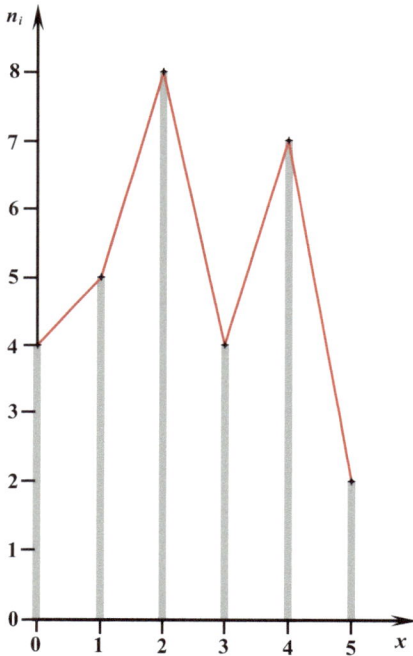

Polygon from a Bars diagram (Table

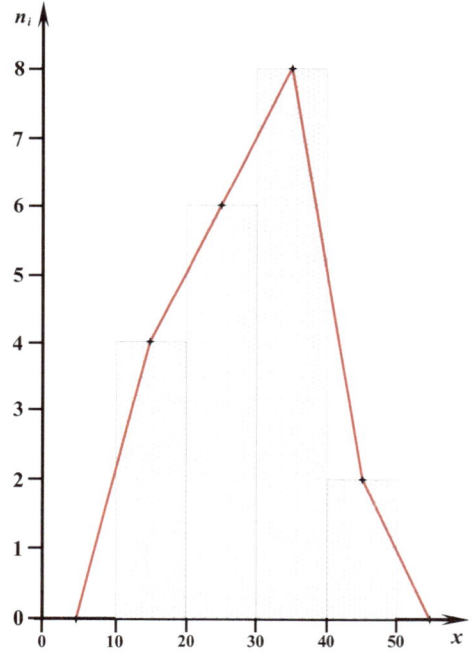

Polygon from Histograms diagram (Table 3)

**Fig. (7.4).** Polygon representation.

**Important note:** *Care* must be taken when graphing a *continuous random variable* with the widths or amplitudes of the classes (intervals). If they are *not equal*, the number of staff (workforce) must be corrected, so that they correspond with *equal widths. For that, we take the class of small amplitude* as reference and we *rectify*.

**Example 03:** We give the following statistical series with a continuous random variable (Table **7.4**):

**Table 7.4. Rectification of the statistical series for unequal classes.**

| $x_i$ | [0-5] | [5-10] | [10-20] | [20-35] | [35-40] | [40-50] |
|---|---|---|---|---|---|---|
| $n_i$ | 4 | 6 | 10 | 12 | 5 | 8 |
| $n_i$ rectified | 4 | 6 | 5 | 4 | 5 | 4 |
| - | - | | 10/2 | 12/3 | - | 8/2 |

We can see that the amplitudes of the classes are unequal. For this, we take the series at amplitude **5** as a reference, because it is the smallest. The series [0-5], [5-10] and [35-40] are all at width **5**, and their staff's number will not undergo any corrections. The other series are larger and corrections are necessary. For example, the series [10-20] has an amplitude 10 (20-10), so, 2×**5** (two times the width of the smallest series). Its staff count is 10, but for the whole series. Consequently, for an amplitude of **5**, we will have the staff's number (Fig. **7.5**). Likewise for the other series (see 3rd and 4th rows of Table **7.3**).

## THE CENTRAL TENDENCY OF A STATISTICAL SERIES AND ITS INDICATORS

In statistics, the *central tendency* term refers to the way in which quantitative data tends to *cluster around a value*. There are several ways to calculate this central value [9, 10, 11, 12].

### The Mode

It represents the *most frequent* statistical variable.

- We can have only one mode  ⇒ *Uni-mode* statistical series (Fig. **7.6.a**);
- We can have several modes  ⇒ *Multi-mode* statistical series (Fig. **7.6.b/c**);
- We may not have modes      ⇒ *Non-modal* statistical series.

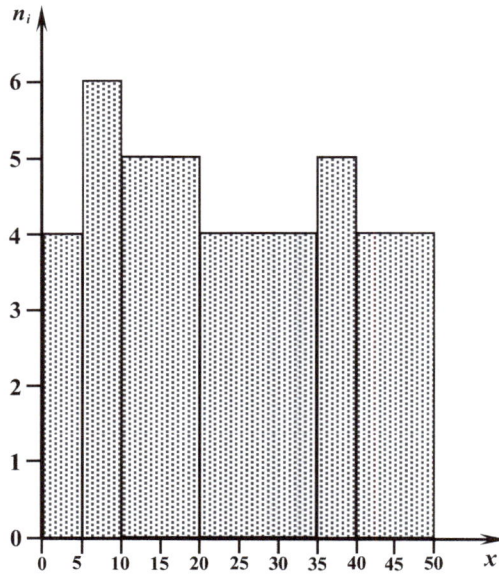

**Fig. (7.5).** Histograms diagram (from Table **7.4**).

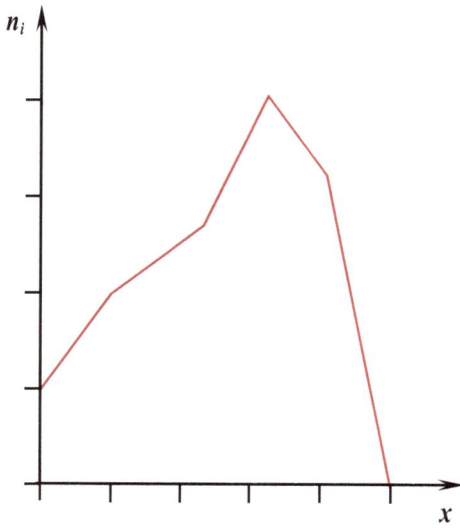

**a.** Uni-mode series

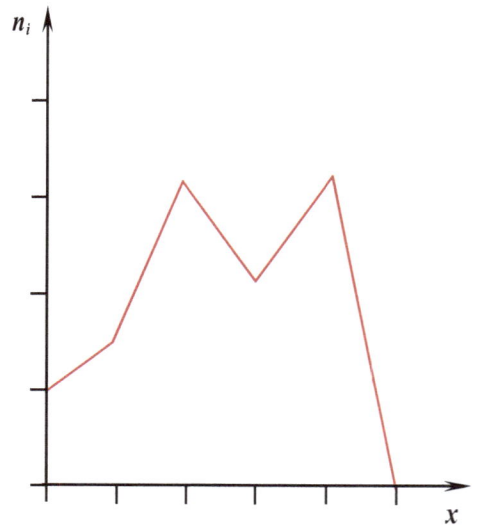

**b.** Bi-modes series

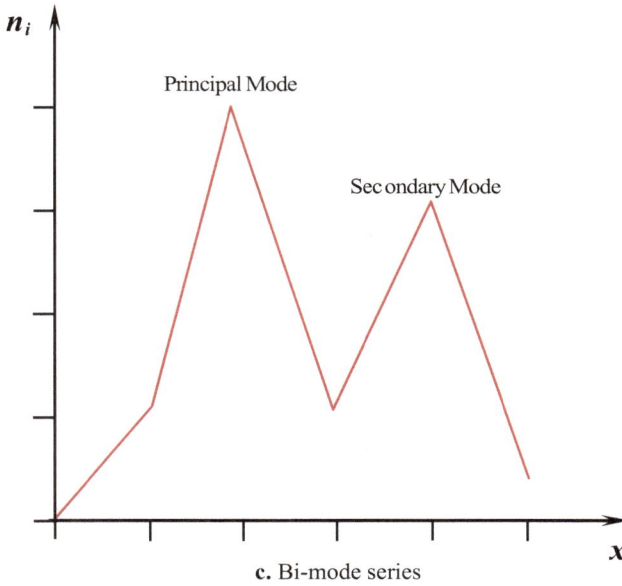

c. Bi-mode series

**Fig. (7.6).** Examples of modal series.

For example 01, the mode is **2**. Since it is repeated 8 times, more times than all the others.

**Example 04:** We have the ages of 20 students: 15 ; 17 ; 18 ; 20 ; 22 ; 24 ; 21 ; 17 ; 16 ; 15 ; 21 ; 22 ; 23 ; 22 ; 17 ; 22 ; 18 ; 22 ; 19 ; 20.

We can put them in an ascending order of Age (Table **7.5**):

**Table 7.5. Mode determination.**

| *Age* | 15 | 16 | 17 | 18 | 19 | 20 | 21 | 22 | 23 | 24 |
|-------|----|----|----|----|----|----|----|----|----|----|
| *Fr*  | 2  | 1  | 3  | 2  | 1  | 2  | 2  | 5  | 1  | 1  |
|       |    |    |    |    |    |    |    |    | the Mode |    |

**Note:** For a *continuous random variable*, the *most frequent class* is said: **modal class**. The *mode* is the *average value* of this class (See the first series of exercises in statistics).

## The Median

It describes the *position of the middle* of a statistical series. There are *as many* values *above* as *below*. Before calculating the median, you *must classify* the values of *x* in *ascending* or *descending order*.

We distinguish *two cases* for a *discrete x*:

-    The number of $x_i$ is odd :

$$Med = x|_{\frac{N+1}{2}} \tag{7.2}$$

-    The number of $x_i$ is even :

$$Med = \frac{1}{2}\left( x|_{\frac{N}{2}} + x|_{\frac{N}{2}+1} \right) \tag{7.3}$$

**Example 05:** We give the series: 7 ; 13 ; 18 ; 10 ; 15.

Ranking in ascending order: 7 ; 10 ; 13 ; 15 ;18.      (Number of $x_i$ odd$\Rightarrow$ $Med = x|_{\frac{5+1}{2}} = x|_3 = 13$ )

We give the series: 7 ; 14 ; 12 ; 8 ; 18 ; 15.

Ranking in ascending order: 7 ; 8 ; 12 ; 14 ; 15 ; 18. (Number of $x_i$ even$\Rightarrow$ $Med = \frac{1}{2}(x|_3 + x|_4) = 13$ )

**Note:** For statistical series without mode, the mode is calculated using the formula:

$$Mod = 3 \times Med - 2 \times \overline{X} \tag{7.4}$$

**The weighted arithmetic mean $\overline{X}$ :** It is calculated by the expression:

$$\overline{X} = \frac{1}{N} \sum_{i=1}^{n} n_i \times x_i \tag{7.5}$$

( $N = \sum_{i=1}^{n} n_i$ : The total workforce)

# Example 06

In a high school, there are four first year classes, containing 25, 26, 30 and 29 students, respectively.

On a common math test, the averages of these classes are 10.5, 11.2, 9.8 and 10.3, respectively.

-   Calculate the average at the control for all the first-year classes.

## Solution

$$\overline{X}_{all\ classes} = \frac{1}{N}\sum_{i=1}^{4} n_i \times x_i = \frac{1}{25+26+30+29}\left(25 \times 10.5 + 26 \times 11.2 + 30 \times 9.8 + 29 \times 10.3\right) = 10.42$$

It is clear that the same procedure was followed to obtain the averages for each class.

## Calculation of the Average from the Frequencies

The average can be calculated from the frequencies without the need to know the number of total staff.

**Example 07:** In a city, 30% of the population have no pets, 40% have one, 15% have two, 10% have three and 5% have four.

-   What is the average number of pets in this city?

## Solution

Here is the calculation to perform:

$$\overline{X} = 0 \times 0.3 + 1 \times 0.4 + 2 \times 0.15 + 3 \times 0.1 + 4 \times 0.05 = 1.2$$

The average is therefore **1.2**. The idea is that each frequency is calculated for the same number of total staff.

**Note:** If the $x_i$ has the same staff's number, the average is said to be arithmetic. Weights (frequencies) have no influence because of their equality.

## STATISTICAL SERIES FORMED ACCORDING TO THE VALUES OF CENTRAL TENDENCY INDICATORS

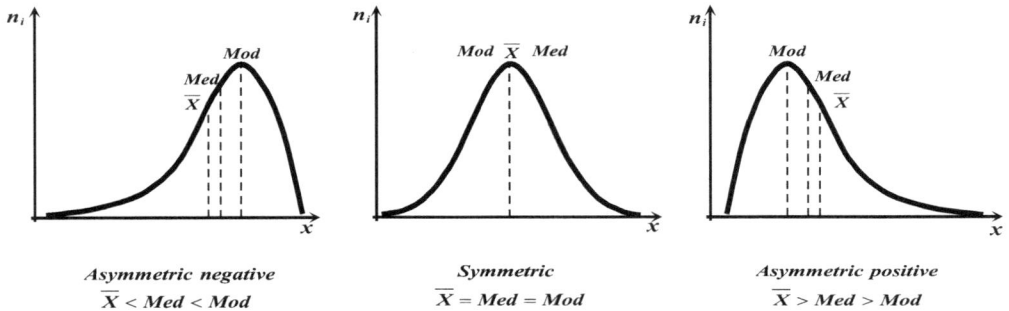

Asymmetric negative
$\overline{X} < Med < Mod$

Symmetric
$\overline{X} = Med = Mod$

Asymmetric positive
$\overline{X} > Med > Mod$

**Fig. (7.7).** Tendency of the statistical series according to its indicators.

## OTHER TYPES OF THE MEAN

### The Harmonic Mean

*The harmonic mean* of *n* numbers, is defined as *n divided by* the sum of the inverses of each number [13-14]. In other words, to calculate the harmonic mean, we *add up the inverse numbers* of each of the observations. Then, we *divide the total number* of observations in our series of values *by the obtained sum*.

$$H = \frac{N}{\sum_{i=1}^{n} n_i / x_i} \tag{7.6}$$

This average is *rarely used*, but it is particularly *useful* in calculating **the rate**.

**Example 08:** A *car driver* moves from town **A** to town **B** with a speed assumed to be constant and equal to $V_{AB} = 80\ Km/h$. His return speed was also assumed to be constant and was $V_{BA} = 20Km/h$.

- What is its *average speed for the entire distance traveled* (outward and return)?

**Solution:** The distance traveled is the same for going and returning. We denote it **D**. Moreover, given the different speeds, the times necessary to cross this distance will be different. We note them: $t_{AB}$ and $t_{BA}$.

So: $D = V_{AB} \times t_{AB}$ ; $D = V_{BA} \times t_{BA}$

The total distance traveled is **2×D** and the total elapsed time is $t_{AB} + t_{BA}$

The average speed is none other than: $V_{av} = \dfrac{2D}{t_{AB} + t_{BA}}$

$$V_{av} = \frac{2}{\dfrac{t_{AB}}{D} + \dfrac{t_{BA}}{D}} = \frac{2}{\dfrac{1}{V_{AB}} + \dfrac{1}{V_{BA}}}$$      **Thus:**      $V_{av} = \dfrac{2}{\dfrac{1}{80} + \dfrac{1}{120}} = 96\,Km/h$

## The Geometric Mean

The *geometric mean* is an instrument for calculating *average rates*, in particular *annual average rates*. Its use *only makes sense* if the values have a *multiplicative character*.

$$\overline{G} = \sqrt[n]{\prod_{i=1}^{n} x_i} \tag{7.7}$$

**Example 09:** The *price* of a product has undergone *annual increases* of **10%** for **07** *years*, then **30%** for **03** *years*.

- What is the *average annual increase rate* during the **10** *years*?

**Solution:** The **10%** increase implies that the price for the *new year* equals **1.1**×the price of the *year just before*. Likewise for **30%**, **Price**$_{ny}$=**1.3**×**Price**$_{yjb}$.

Therefore, the initial price ($1^{st}$ year) underwent the following increase after 10 years:

$$\text{Price}_{10^{th}\,y} = \text{Price}_{1^{st}\,y} \times \left(1.1 \times 1.1 \times 1.1 \times 1.1 \times 1.1 \times 1.1 \times 1.1 \times 1.3 \times 1.3 \times 1.3\right)$$
$$= \text{Price}_{1^{st}\,y} \times \left(1.1^7 \times 1.3^3\right)$$

Here, the price was calculated at *the end* of the **10**<sup>th</sup> *year*.

The average increase rate is: $\text{IR}_{av}{}^{10y} = 1.1^7 \times 1.3^3 \Rightarrow \text{IR}_{av} = \sqrt[10]{1.1^7 \times 1.3^3} \approx 1.1565 = \overline{G}$

## *The Quadratic Mean*

*It is an average* that finds applications when one is dealing with phenomena presenting a *sinusoidal character*, with alternating positive and negative values. It is, therefore, *widely used in electricity*. It makes it possible in particular to calculate the magnitude of a set of numbers.

$$\overline{Q} = \sqrt{\frac{1}{n}\sum_{i=1}^{n} x_i} \qquad (7.8)$$

**Example 10:** We assume three squares with sides **5**cm, 7cm and **9**cm, respectively.

- Do you think the average air of the three squares is equal to the area of a square with side 7 Cm?

**Solution:** We know that the area of a square is the square of its side. Then the average area of the three squares is calculated as follows:

$$A_{av} = \frac{1}{3}\left( Sq_1^2 + Sq_2^2 + Sq_3^2 \right) = \frac{1}{3}\left( 5^2 + 7^2 + 9^2 \right) \approx 51.667\,cm^2$$

Then, the side of the average square area will be: $Si_{av} = \sqrt{A_{av}} \approx 7.188\,cm$

- So, the *answer* is **no** for the square of **7**cm.

Here, we can see that the arithmetic mean does not give the right result but the quadratic mean does.

**Important Notes:**

- For the averages detailed above, we can associate the weighting. i.e. the number $n_i$ for each value $x_i$, as done for the weighted arithmetic mean;
- When working with statistical classes (continuous variable defined on intervals), the laws of averages remain usable. But, we must take the value at the interval center as the value of $x_i$. Review the polygon plot for a continuous statistical variable (Fig. **7.4**).

**The Extent**

It is the *difference between* the **highest** and **lowest** values for a statistical series. Its value is *greatly influenced* by the presence of an *outlier* in the data (very small or very large). Before calculating it, it is necessary (to avoid any error) to order the data in increase or decrease order.

**Example 11:** We give the series: 7 ; 13 ; 18 ; 10 ; 15.

Ranking by growth: 7 ; 10 ; 13 ; 15 ; 18.

The extent is: $Ext=18-7=11$.

## Quartiles

Their nomination comes from the quarter (1/4). *Dividing* the *ordered* statistical series into **four** pieces of *equal extents* (25% of the total extent), gives three limits (in addition to the minimum and maximum values of course). The first named $Q_1$, the first quartile. The second $Q_2$, the second quartile or the **median** (**Med**), and the third $Q_3$, the third quartile (Fig. **7.8**).

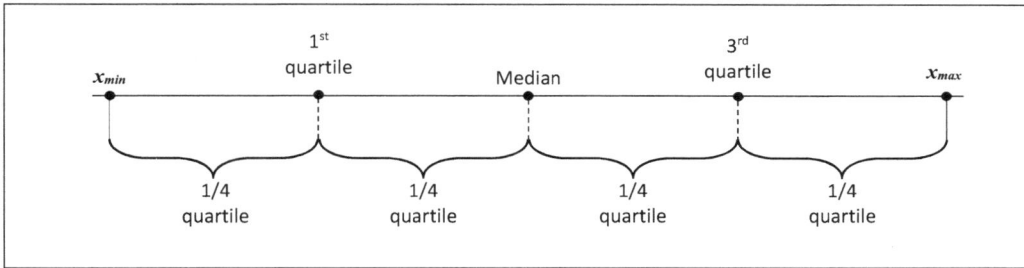

**Fig. (7.8).** Schematization of quartiles.

Among the purposes of calculating quartiles is the exclusion (under reservation) of small and large values between $x_{min}$ and $Q_1$ and between $Q_3$ and $x_{max}$. *The majority* (50%) is located between $Q_1$ and $Q_3$, and will *receive more interest. This can be understood* when the series *has a central tendency*, where great efforts will be made without much interest when one is interested in small and large values, *especially for very numerous* statistical series.

## Notes:

- The rules for series with an *even* or *odd* number of variables remain applicable for quartiles.
- $Q_3-Q_1$ is called the interquartile range;
- When $(Q_3-Q_1)/Ext<0.5$, we say that there is a strong central concentration of the series. Extreme values are losing their influence. Otherwise $(Q_3-Q_1)/Ext>0.5$, we have a strong central dispersion, and we can no longer neglect the effects of extreme values;
- **Deciles** and **percentiles** are sometimes defined when the *series* is *very dispersed*, or with a *very large number of* $x_i$. The **deciles** are for *dividing* the

series into *ten pieces*, while the **percentiles** are for *dividing* into *one hundred pieces*. *The details of these parameters are beyond the scope of our course.*

## STANDARD DEVIATION

*The standard deviation* is an *indicator of dispersion* (*as seen in* the *probability part*) (Fig. **7.9**) [8, 9, 13]. It informs us about the way in which individuals are distributed around the average. It gives the value of the average deviation from the arithmetic mean (weighted or not). It is calculated with one of the following formulas:

$$\sigma = \sqrt{V} = \sqrt{\frac{1}{N}\sum_{i=1}^{n} n_i \left( x_i - \overline{X} \right)^2} \quad \text{Or:} \quad \sigma = \sqrt{V} = \sqrt{\frac{1}{N}\sum_{i=1}^{n} n_i . x_i^2 - \overline{X}^2} \qquad (7.9)$$

The term $V$ is called *the variance*.

-In most cases, the standard deviation is an excellent indicator of the dispersion of the $x_i$ around the mean. It is *influenced* by *outliers*, when very large values are located at the ends for example.

-   A **small** *standard deviation* indicates that most of the $x_i$ are close to $\overline{X}$ , and we speak of a *condensed* or *little-dispersed* statistical series;
-   A **large** *standard deviation* indicates that most of the $x_i$ are far from $\overline{X}$ , and we speak of a highly dispersed statistical series;
-   Note that the standard deviation is *always positive* (or *nil*)

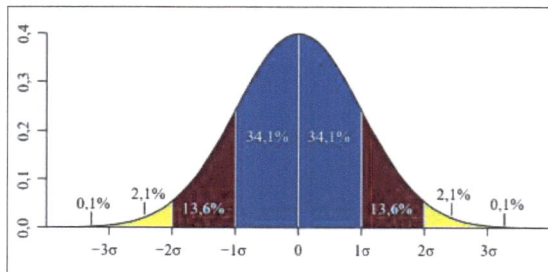

**Fig. (7.9).** Graphical representation of the normal distribution 'density function. Each colored band is one standard deviation wide.

Approximately (case of outlier $x_i$ excluded):

→ **68%** of values are in the range: $\overline{X} - \sigma \le x_i \le \overline{X} + \sigma$

→ **95%** of values are in the range: $\overline{X} - 2\sigma \le x_i \le \overline{X} + 2\sigma$

→ **99%** of values are in the range: $\overline{X} - 3\sigma \le x_i \le \overline{X} + 3\sigma$

## Coefficient of Variation (or Relative Standard Deviation)

The coefficient of variation is *a relative measure of the dispersion* of the data around the mean. The coefficient of variation is calculated as the ratio of the standard deviation to the mean and is expressed as a percentage. It gives more accuracy on the dispersion compared to the standard deviation.

It is defined by:

$$CV = \frac{\sigma}{\overline{X}} \tag{7.10}$$

-   $CV < 0.5$ (*or* $< 50\%$) : Low dispersion;
-   $CV > 0.5$ (*or* $> 50\%$) : Strong dispersion;

**Example 12:** The weights of **300** *workers* in a company are given in the form of a histogram.

- Determine the standard deviation of this series and criticize the weights dispersion.

## Solution

The calculation of the standard deviation must go through the calculation of the mean. The latter, for a continuous statistical series, is calculated using the centers of each class.

We will plot the data and the stages of calculation in the Table (**7.6**).

**Table 7.6. Worker Statistics and calculations.**

| $w_i$ (Kg) | [62-66] | [66-70] | [70-74] | [74-78] | [78-82] | [82-86] | [86-90] | - |
|---|---|---|---|---|---|---|---|---|
| $n_i$ ($N^{br}$ work) | 13 | 40 | 96 | 65 | 30 | 50 | 6 | *N=300* |
| $c_i$ (class center) | 64 | 68 | 72 | 76 | 80 | 84 | 88 | - |
| *Fr* | 13/300 | 40/300 | 96/300 | 65/300 | 30/300 | 50/300 | 6/300 | - |
| *Fr* × $c_i$ | $\frac{832}{300}$ | $\frac{2720}{300}$ | $\frac{6912}{300}$ | $\frac{4940}{300}$ | $\frac{2400}{300}$ | $\frac{4200}{300}$ | $\frac{528}{300}$ | - |
| *Fr*.$c_i^2$ | $\frac{53248}{300}$ | $\frac{184960}{300}$ | $\frac{497664}{300}$ | $\frac{375440}{300}$ | $\frac{192000}{300}$ | $\frac{352800}{300}$ | $\frac{46464}{300}$ | - |

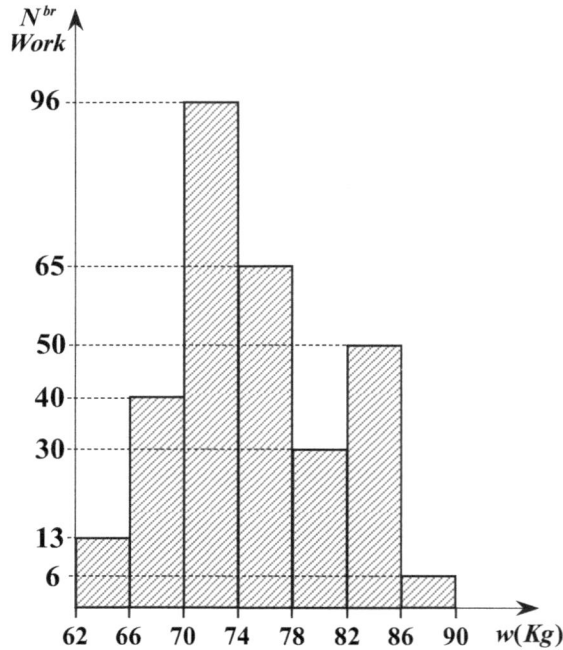

**Fig. (7.10).** Histogram of the company workers' weights.

From the Table:

$$\overline{X} = \sum_{i=1}^{7} n_i.c_i = 22532/300 \simeq 75.10 \, Kg$$

$$\sigma = \sqrt{\sum_{i=1}^{7} Fr_i.c_i^2 - \overline{X}^2} = \sqrt{\frac{1702576}{300} - 5641.0114} \simeq 5.85 \, Kg$$

To criticize the dispersion, we have to calculate the variation coefficient:
$CV = \sigma/\overline{X} = 5.85/75.10 \simeq 7.79\%$

The *CV* value is much less than **50%**. We can say the degree of dispersion is very small. This indicates a *concentration of workers' weights around the average,* recalling that **68%** of *the workers' weight* is between **69.25Kg** and **80.95Kg**.

## CONFIDENCE INTERVAL

As we detailed at the beginning of this part, when the population is *large* (very *high staff' number*) or we have difficulties in listing the information of all individuals, we take *a sample*, which *will serve* as *an indicator* for the *whole population* [9, 11-14]. Of course, the *number* of individuals in the sample and the *diversity are two important parameters* to consider. But, because there may be differences in the entire population compared to the sample, specialists have developed techniques to be more confidential when spreading the statistical results of the sample over the entire population.

There are **several cases** for the situation. We will *limit ourselves* to *the case* where the **number of individuals** in the sample is **greater than 30** (**n>30**), a *condition required to be able* to assume that our distribution can be *approximated by a **normal distribution*** (review the last section of the *probability part*). The *advantage* of this requirement is that the use of the **Z-table** of the standard normal distribution is *very easy and motivating*. There are also *several cases*, where the population mean and its standard deviation are known or not, both at the same time or independently. In the following, we will take two classic cases to illustrate the methodology for determining the confidence interval for the mean and the standard deviation of the population. *The cases where **n≤30** will also be presented* but with a *single example* to show how to use the ***t-Student table***.

The **Z-table** *is shown below*. It should be remembered that *this table* is for *a mean equal to* **1** with a *zero standard deviation* ( $\aleph(0,1)$ ), which are the two parameters of the *standard* normal law. **Any normal distribution** *can become* **standard** by changing a variable $Z = (x - \mu)/\sigma$ (hence the name **Z-table**), with **x being** the value of our *random statistical variable*, **μ**: the *population mean* and **σ**: its *standard deviation*. It is obvious that the *normal* (*or standard*) *law* has *symmetry* with respect to the passing vertical by **x=μ**. For this, we can find a **Z-table** on the *left* (with signs -) or on the *right* (with signs +). It is necessary to **be careful** *during the calculation only*.

## Risk of Error

In a simple way, to specify the confidence interval, we must know the degree of error made. This error is noted **α**. Then, the confidence interval will be covered by the probability area $\int_{-Z_{\alpha/2}}^{+Z_{\alpha/2}} f(Z).dZ = 1 - \alpha$. The symmetry of the normal distribution leads to place the error on both sides of the distribution curve with **α/2** as area (Fig. **7.11**).

Although the value of **α** is fixed by the statistician by choice in general, but the most used values are: **10%, 5%, 1%** and **0.1%**. This leads to confidence intervals of: **90%, 95%, 99%** and **99.9%** respectively (Table **7.7**).

**Table 7.7. Z-Table.**

| z | 0.00 | 0.01 | 0.02 | 0.03 | 0.04 | 0.05 | 0.06 | 0.07 | 0.08 | 0.09 |
|-----|--------|--------|--------|--------|--------|--------|--------|--------|--------|--------|
| 0.0 | 0.0000 | 0.0040 | 0.0080 | 0.0120 | 0.0160 | 0.0199 | 0.0239 | 0.0279 | 0.0319 | 0.0359 |
| 0.1 | 0.0398 | 0.0438 | 0.0478 | 0.0517 | 0.0557 | 0.0596 | 0.0636 | 0.0675 | 0.0714 | 0.0753 |
| 0.2 | 0.0793 | 0.0832 | 0.0871 | 0.0910 | 0.0948 | 0.0987 | 0.1026 | 0.1064 | 0.1103 | 0.1141 |
| 0.3 | 0.1179 | 0.1217 | 0.1255 | 0.1293 | 0.1331 | 0.1368 | 0.1406 | 0.1443 | 0.1480 | 0.1517 |
| 0.4 | 0.1554 | 0.1591 | 0.1628 | 0.1664 | 0.1700 | 0.1736 | 0.1772 | 0.1808 | 0.1844 | 0.1879 |
| 0.5 | 0.1915 | 0.1950 | 0.1985 | 0.2019 | 0.2054 | 0.2088 | 0.2123 | 0.2157 | 0.2190 | 0.2224 |
| 0.6 | 0.2257 | 0.2291 | 0.2324 | 0.2357 | 0.2389 | 0.2422 | 0.2454 | 0.2486 | 0.2517 | 0.2549 |
| 0.7 | 0.2580 | 0.2611 | 0.2642 | 0.2673 | 0.2704 | 0.2734 | 0.2764 | 0.2794 | 0.2823 | 0.2852 |
| 0.8 | 0.2881 | 0.2910 | 0.2939 | 0.2967 | 0.2995 | 0.3023 | 0.3051 | 0.3078 | 0.3106 | 0.3133 |
| 0.9 | 0.3159 | 0.3186 | 0.3212 | 0.3238 | 0.3264 | 0.3289 | 0.3315 | 0.3340 | 0.3365 | 0.3389 |
| 1.0 | 0.3413 | 0.3438 | 0.3461 | 0.3485 | 0.3508 | 0.3531 | 0.3554 | 0.3577 | 0.3599 | 0.3621 |
| 1.1 | 0.3643 | 0.3665 | 0.3686 | 0.3708 | 0.3729 | 0.3749 | 0.3770 | 0.3790 | 0.3810 | 0.3830 |
| 1.2 | 0.3849 | 0.3869 | 0.3888 | 0.3907 | 0.3925 | 0.3944 | 0.3962 | 0.3980 | 0.3997 | 0.4015 |
| 1.3 | 0.4032 | 0.4049 | 0.4066 | 0.4082 | 0.4099 | 0.4115 | 0.4131 | 0.4147 | 0.4162 | 0.4177 |
| 1.4 | 0.4192 | 0.4207 | 0.4222 | 0.4236 | 0.4251 | 0.4265 | 0.4279 | 0.4292 | 0.4306 | 0.4319 |
| 1.5 | 0.4332 | 0.4345 | 0.4357 | 0.4370 | 0.4382 | 0.4394 | 0.4406 | 0.4418 | 0.4429 | 0.4441 |
| 1.6 | 0.4452 | 0.4463 | 0.4474 | 0.4484 | 0.4495 | 0.4505 | 0.4515 | 0.4525 | 0.4535 | 0.4545 |
| 1.7 | 0.4554 | 0.4564 | 0.4573 | 0.4582 | 0.4591 | 0.4599 | 0.4608 | 0.4616 | 0.4625 | 0.4633 |
| 1.8 | 0.4641 | 0.4649 | 0.4656 | 0.4664 | 0.4671 | 0.4678 | 0.4686 | 0.4693 | 0.4699 | 0.4706 |
| 1.9 | 0.4713 | 0.4719 | 0.4726 | 0.4732 | 0.4738 | 0.4744 | 0.4750 | 0.4756 | 0.4761 | 0.4767 |
| 2.0 | 0.4772 | 0.4778 | 0.4783 | 0.4788 | 0.4793 | 0.4798 | 0.4803 | 0.4808 | 0.4812 | 0.4817 |
| 2.1 | 0.4821 | 0.4826 | 0.4830 | 0.4834 | 0.4838 | 0.4842 | 0.4846 | 0.4850 | 0.4854 | 0.4857 |
| 2.2 | 0.4861 | 0.4864 | 0.4868 | 0.4871 | 0.4875 | 0.4878 | 0.4881 | 0.4884 | 0.4887 | 0.4890 |
| 2.3 | 0.4893 | 0.4896 | 0.4898 | 0.4901 | 0.4904 | 0.4906 | 0.4909 | 0.4911 | 0.4913 | 0.4916 |
| 2.4 | 0.4918 | 0.4920 | 0.4922 | 0.4925 | 0.4927 | 0.4929 | 0.4931 | 0.4932 | 0.4934 | 0.4936 |
| 2.5 | 0.4938 | 0.4940 | 0.4941 | 0.4943 | 0.4945 | 0.4946 | 0.4948 | 0.4949 | 0.4951 | 0.4952 |
| 2.6 | 0.4953 | 0.4955 | 0.4956 | 0.4957 | 0.4959 | 0.4960 | 0.4961 | 0.4962 | 0.4963 | 0.4964 |
| 2.7 | 0.4965 | 0.4966 | 0.4967 | 0.4968 | 0.4969 | 0.4970 | 0.4971 | 0.4972 | 0.4973 | 0.4974 |
| 2.8 | 0.4974 | 0.4975 | 0.4976 | 0.4977 | 0.4977 | 0.4978 | 0.4979 | 0.4979 | 0.4980 | 0.4981 |
| 2.9 | 0.4981 | 0.4982 | 0.4982 | 0.4983 | 0.4984 | 0.4984 | 0.4985 | 0.4985 | 0.4986 | 0.4986 |
| 3.0 | 0.4987 | 0.4987 | 0.4987 | 0.4988 | 0.4988 | 0.4989 | 0.4989 | 0.4989 | 0.4990 | 0.4990 |

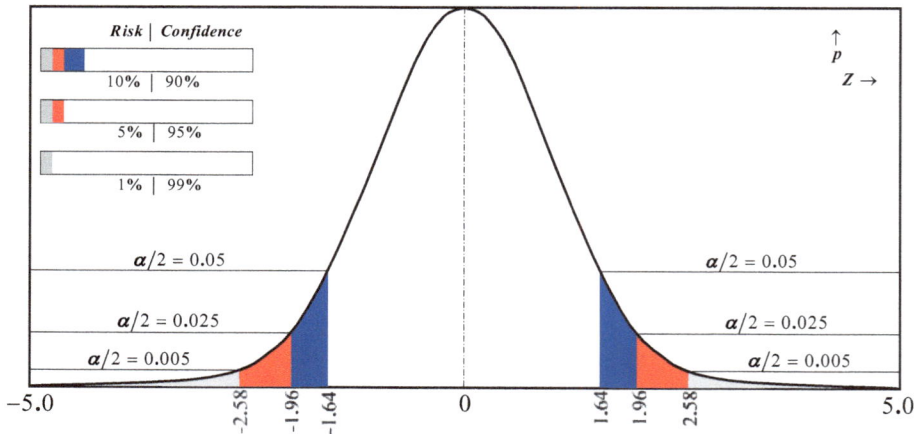

**Fig. (7.11).** Risk and confidence areas on a normal distribution.

The values of **Z** are determined from the **Z-Table** by looking for the value $(1-\alpha)/2$, inside the table, then *we read* the corresponding values on *the left* (1st column) and at *the top* (1st row), which we *will sum them* to have the *corresponding value of* **Z**.

For example, in the **Z-Table** *above*, the <u>circled value</u> equals **0.4750**. Therefore: $(1-\alpha)/2 = 0.475 \Rightarrow \alpha = 0.05$. Thus: **Z**=1.90+0.06=**1.96**. This is the value corresponding to a **95%** *confidence level*. The same goes for other levels of confidence.

As:

$$Z = \frac{\overline{X} - \mu}{\sqrt{\sigma^2/n}} \tag{7.11}$$

We can determine the *limit values* (*lower* and *upper*) of the population mean ($\mu$) from the *sample mean, according to the chosen confidence level.*

$$-Z_{\alpha/2} \leq Z = \frac{\overline{X} - \mu}{\sqrt{\sigma^2/n}} \leq Z_{\alpha/2} \tag{7.12}$$

$$\Rightarrow \quad \boxed{\overline{X} - Z_{\alpha/2}\sqrt{\sigma^2/n} \leq \mu \leq \overline{X} + Z_{\alpha/2}\sqrt{\sigma^2/n}} \tag{7.13}$$

**Note:** In these expressions, we find **n**, which represents the *number of sample staff* (this is when we apply the laws of probability in statistics). <u>*The demonstration of these formulas is long and goes beyond the scope of this course*</u>.

Several situations can be encountered. We are going to present *just two of them only*.

**Case Where the Variance (and Standard Deviation) of the Population is Known ($\sigma^2$)**

**Example 13**: If you are informed that the *variance in weight* of the university students equals **144Kg²,** and we took **100** *students* as a sample and the *average weight* of this sample is **64Kg**.

- Estimate from these data the average weight of all the students with a confidence level equal to **99%**, then with **95%**.

**Solution:** In this example, *the variance* of the population *is known*. We will determine the lower and upper limits of the population weight from the last expression and use the **Z-table** to get the value of $Z_{\alpha/2}$.

o   The confidence level is assumed to be equal to **0.99**.

Therefore: $1-\alpha = 0.99 \Rightarrow \dfrac{1-\alpha}{2} = \dfrac{0.99}{2} = 0.495$

From the **Z-table**:   $Z_{\alpha/2} \simeq 2.58$

Consequently:

$$\mu_{L\inf} = \overline{X} - Z_{\alpha/2}\sqrt{\sigma^2/n}$$
$$= 64 - 2.58 \times \sqrt{144/100} = 64 - 2.58 \times 1.2 \simeq 60.90 \textbf{\textit{Kg}}$$

$$\mu_{L\sup} = \overline{X} + Z_{\alpha/2}\sqrt{\sigma^2/n}$$
$$= 64 + 2.58 \times \sqrt{144/100} = 64 + 2.58 \times 1.2 \simeq 67.09 \textbf{\textit{Kg}}$$

So, we are **99%** certain that the average university weight is between **60.90***Kg* and **67.09***Kg*.

o   The confidence level is assumed to be equal to **0.95.**

$$\text{For } 1-\alpha = 0.95 \Rightarrow \dfrac{1-\alpha}{2} = \dfrac{0.95}{2} = 0.475$$

From the Z-*table*: :   $Z_{\alpha/2} \simeq 1.96$

$$\mu_{L\inf} = 64 - 1.96 \times \sqrt{144/100} = 64 - 1.96 \times 1.2 \simeq 61.65 \textbf{\textit{Kg}}$$

$$\mu_{L\sup} = 64 + 1.96 \times \sqrt{144/100} = 64 + 1.96 \times 1.2 \simeq 66.35 \textbf{\textit{Kg}}$$

So, we are **95%** certain that the average university weight is between **61.65***Kg* and **66.35***Kg*.

**Case Where the Variance (Or Standard Deviation) of the Population is Unknown ($\sigma^2 = ?$)**

For this situation, *there are two cases*:

**b.1. $n>30$** : We use *Z-table* and we take the value of the variance of the sample, usually we name it $S^2$ in the expression of $\mu$.

Where : $S^2$ is the variance of the sample, calculated as follows:

$$S^2 = \frac{1}{(n-1)} \sum_{i=1}^{n} x_i^2 - \overline{X}^2 \qquad (7.14)$$

$(n-1) = \upsilon :$ $N^{br}$ *dof* (degrees of freedom)

**b.2. $n\leq30$** : In this case, instead of $Z_{\alpha/2}$ we utilize $t_{\alpha/2}$. The expression for $\mu$ remains the same. The values of $t_{\alpha/2}$ (depending on the confidence level) are taken from the *t-Student* table (Table **7.8**). We can see that the values converge towards those of $Z_{\alpha/2}$ when *n approaches* **30**.

It should be noted that *t* is a variable change (as *Z*), with the expression:

$$t = \frac{\overline{X} - \mu}{\sqrt{S^2/n}} \qquad (7.15)$$

**Example 14:** If you are informed that the weight of boxes of Spaghetti, manufactured by a factory, follows a *normal law*. **25** boxes were *randomly drawn*. After measurements and calculation, we found that the *average weight* is **498.25g** with a *variance* of **1.69g²**.

-We want to know the *average weight* for *all boxes* made with a **90%** *confidence level*.

## Solution

We can see that the *variance* of the <u>population</u> is *not known*. In addition, the *sample size* is *less than* **30**. So, we will use the expression for $\mu$ with the precisions in **b.2**.

We have: $\upsilon = n-1 = 25-1 = 24$ ; $\overline{X} = 498.25$ ; $S^2 = 1.69$ ;

$$(1-\alpha) = 0.9 \Rightarrow (\alpha/2) = 0.05$$

From *t-Student table*: $t_{\alpha/2} = 1.711$

Therefore:

$$\mu_{Linf} = 498.25 - 1.711 \times \sqrt{1.69/25} \approx 497.80g$$

$$\mu_{Lsup} = 498.25 + 1.711 \times \sqrt{1.69/25} \approx 498.69g$$

We are **90%** sure that the weight of the boxes varies between **497.80**g and **498.69**g.

**Table 7.8.** *T-Student* table (usable for $n \leq 30$).

| | | | | | $\alpha/2$ | | | | | |
|---|---|---|---|---|---|---|---|---|---|---|
| $\upsilon$ | 0.40 | 0.25 | 0.10 | 0.05 | 0.025 | 0.01 | 0.005 | 0.0025 | 0.001 | 0.0005 |
| 1 | 0.325 | 1.000 | 3.078 | 6.314 | 12.706 | 31.821 | 63.657 | 127.321 | 318.309 | 636.619 |
| 2 | 0.289 | 0.816 | 1.886 | 2.920 | 4.303 | 6.965 | 9.925 | 14.089 | 22.327 | 31.599 |
| 3 | 0.277 | 0.765 | 1.638 | 2.353 | 3.182 | 4.541 | 5.841 | 7.453 | 10.215 | 12.924 |
| 4 | 0.271 | 0.741 | 1.533 | 2.132 | 2.776 | 3.747 | 4.604 | 5.598 | 7.173 | 8.610 |
| 5 | 0.267 | 0.727 | 1.476 | 2.015 | 2.571 | 3.365 | 4.032 | 4.773 | 5.893 | 6.869 |
| 6 | 0.265 | 0.718 | 1.440 | 1.943 | 2.447 | 3.143 | 3.707 | 4.317 | 5.208 | 5.959 |
| 7 | 0.263 | 0.711 | 1.415 | 1.895 | 2.365 | 2.998 | 3.499 | 4.029 | 4.785 | 5.408 |
| 8 | 0.262 | 0.706 | 1.397 | 1.860 | 2.306 | 2.896 | 3.355 | 3.833 | 4.501 | 5.041 |
| 9 | 0.261 | 0.703 | 1.383 | 1.833 | 2.262 | 2.821 | 3.250 | 3.690 | 4.297 | 4.781 |
| 10 | 0.260 | 0.700 | 1.372 | 1.812 | 2.228 | 2.764 | 3.169 | 3.581 | 4.144 | 4.587 |
| 11 | 0.260 | 0.697 | 1.363 | 1.796 | 2.201 | 2.718 | 3.106 | 3.497 | 4.025 | 4.437 |
| 12 | 0.259 | 0.695 | 1.356 | 1.782 | 2.179 | 2.681 | 3.055 | 3.428 | 3.930 | 4.318 |
| 13 | 0.259 | 0.694 | 1.350 | 1.771 | 2.160 | 2.650 | 3.012 | 3.372 | 3.852 | 4.221 |
| 14 | 0.258 | 0.692 | 1.345 | 1.761 | 2.145 | 2.624 | 2.977 | 3.326 | 3.787 | 4.140 |
| 15 | 0.258 | 0.691 | 1.341 | 1.753 | 2.131 | 2.602 | 2.947 | 3.286 | 3.733 | 4.073 |
| 16 | 0.258 | 0.690 | 1.337 | 1.746 | 2.120 | 2.583 | 2.921 | 3.252 | 3.686 | 4.015 |
| 17 | 0.257 | 0.689 | 1.333 | 1.740 | 2.110 | 2.567 | 2.898 | 3.222 | 3.646 | 3.965 |
| 18 | 0.257 | 0.688 | 1.330 | 1.734 | 2.101 | 2.552 | 2.878 | 3.197 | 3.610 | 3.922 |
| 19 | 0.257 | 0.688 | 1.328 | 1.729 | 2.093 | 2.539 | 2.861 | 3.174 | 3.579 | 3.883 |
| 20 | 0.257 | 0.687 | 1.325 | 1.725 | 2.086 | 2.528 | 2.845 | 3.153 | 3.552 | 3.850 |
| 21 | 0.257 | 0.686 | 1.323 | 1.721 | 2.080 | 2.518 | 2.831 | 3.135 | 3.527 | 3.819 |
| 22 | 0.256 | 0.686 | 1.321 | 1.717 | 2.074 | 2.508 | 2.819 | 3.119 | 3.505 | 3.792 |
| 23 | 0.256 | 0.685 | 1.319 | 1.714 | 2.069 | 2.500 | 2.807 | 3.104 | 3.485 | 3.768 |
| 24 | 0.256 | 0.685 | 1.318 | 1.711 | 2.064 | 2.492 | 2.797 | 3.091 | 3.467 | 3.745 |
| 25 | 0.256 | 0.684 | 1.316 | 1.708 | 2.060 | 2.485 | 2.787 | 3.078 | 3.450 | 3.725 |
| 26 | 0.256 | 0.684 | 1.315 | 1.706 | 2.056 | 2.479 | 2.779 | 3.067 | 3.435 | 3.707 |
| 27 | 0.256 | 0.684 | 1.314 | 1.703 | 2.052 | 2.473 | 2.771 | 3.057 | 3.421 | 3.690 |
| 28 | 0.256 | 0.683 | 1.313 | 1.701 | 2.048 | 2.467 | 2.763 | 3.047 | 3.408 | 3.674 |
| 29 | 0.256 | 0.683 | 1.311 | 1.699 | 2.045 | 2.462 | 2.756 | 3.038 | 3.396 | 3.659 |
| 30 | 0.256 | 0.683 | 1.310 | 1.697 | 2.042 | 2.457 | 2.750 | 3.030 | 3.385 | 3.646 |
| 31 | 0.256 | 0.682 | 1.309 | 1.696 | 2.040 | 2.453 | 2.744 | 3.022 | 3.375 | 3.633 |
| 32 | 0.255 | 0.682 | 1.309 | 1.694 | 2.037 | 2.449 | 2.738 | 3.015 | 3.365 | 3.622 |
| 33 | 0.255 | 0.682 | 1.308 | 1.692 | 2.035 | 2.445 | 2.733 | 3.008 | 3.356 | 3.611 |
| 34 | 0.255 | 0.682 | 1.307 | 1.691 | 2.032 | 2.441 | 2.728 | 3.002 | 3.348 | 3.601 |
| 35 | 0.255 | 0.682 | 1.306 | 1.690 | 2.030 | 2.438 | 2.724 | 2.996 | 3.340 | 3.591 |
| 36 | 0.255 | 0.681 | 1.306 | 1.688 | 2.028 | 2.434 | 2.719 | 2.990 | 3.333 | 3.582 |
| 37 | 0.255 | 0.681 | 1.305 | 1.687 | 2.026 | 2.431 | 2.715 | 2.985 | 3.326 | 3.574 |
| 38 | 0.255 | 0.681 | 1.304 | 1.686 | 2.024 | 2.429 | 2.712 | 2.980 | 3.319 | 3.566 |
| 39 | 0.255 | 0.681 | 1.304 | 1.685 | 2.023 | 2.426 | 2.708 | 2.976 | 3.313 | 3.558 |
| 40 | 0.255 | 0.681 | 1.303 | 1.684 | 2.021 | 2.423 | 2.704 | 2.971 | 3.307 | 3.551 |
| 50 | 0.255 | 0.679 | 1.299 | 1.676 | 2.009 | 2.403 | 2.678 | 2.937 | 3.261 | 3.496 |
| 60 | 0.254 | 0.679 | 1.296 | 1.671 | 2.000 | 2.390 | 2.660 | 2.915 | 3.232 | 3.460 |
| 80 | 0.254 | 0.678 | 1.292 | 1.664 | 1.990 | 2.374 | 2.639 | 2.887 | 3.195 | 3.416 |
| 100 | 0.254 | 0.677 | 1.290 | 1.660 | 1.984 | 2.364 | 2.626 | 2.871 | 3.174 | 3.390 |
| 200 | 0.254 | 0.676 | 1.286 | 1.653 | 1.972 | 2.345 | 2.601 | 2.839 | 3.131 | 3.340 |
| $z$ | 0.253 | 0.674 | 1.282 | 1.645 | 1.960 | 2.326 | 2.576 | 2.807 | 3.090 | 3.291 |
| | 20% | 50% | 80% | 90% | 95% | 98% | 99% | 99.5% | 99.8% | 99.9% |
| | | | | | | $1 - \alpha$ | | | | | |

# Series of Probability Exercises

**Series N° 01**

**Sets**

**Exercise 01**

Determine $x$ and $y$ so that **I = J**, in the following cases [1-3]:

- **I** = {5; 2; 8; 6; 9}　and　**J** = {5; 8; $x$}
- **I** = {k; b; $x$; a; c}　and　**J** = {b; d; $y$; h; a}
- **I** = {19; $x$; $y$}　　　and　**J** = {24; 20; 7; 25; 19}

Note: $x$ and $y$ can be more than one element (subset).

**Exercise 02**

Draw a VENN diagram that represents the set: **A**={2; 4; 7; 9; 13} and place the numbers **0** and **20** on this diagram.

-　Complete the VENN diagram if we have both sets:

**B**={1 ; 3 ; 4 ; 6 ; 7 ; 11}　　　**C**={4 ; 6 ; 9 ; 10}

**Exercise 03**

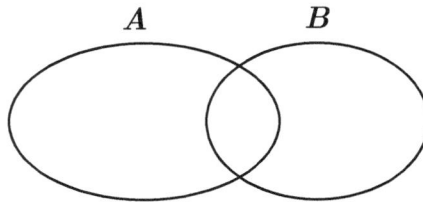

On the following diagram, hatch the empty parts:

-　If **A ⊂ B**

-　If **B ⊂ A**

-　If **A = B**

# Exercise 04

And let three subsets of **S** be defined as follows:

Let the set **S**={0; 1; 2; 3; 4; 5; 6; 7; 8; 9}

**A**={1 ; 2 ; 3 ; 4 ; 5 ; 7}          **B**={*x/x* a number>4}          **C**={*x/x* an even number}

- Determine the following sets:  $B \cap C$ ; $A \cup B \cap C$ ; $A \cup B \cap A \cup C$ ; $A^c$ ; $C^c$ ; $A^c \cap C^c$ ; $A \cup C^{c}$ ; $A^c \cup C^c$ ; $A \cap C^{c}$
- What do you notice?

# Exercise 05

Out of **50 people**, **24** practice *Tennis* and **15** *Swimming*; **06 people** practice *both sports*.

- How many people do not practice any sport?
- We question a person at random. What are the probabilities of having
- Choose a person practicing *only one sport*.
- Choose a person who plays *Tennis*.

# Exercise 06
A statistic made on **100 people** gave: **75** *people* have already owned a *European brand car*; **60** *people* have already owned an *Asian brand car* and **45** have already owned a *European* and *Asian car*.
- How many of the people questioned who *did not own* a car (neither European nor Asian)?
- How many of the people questioned those who have already owned a European car but not an Asian?
- Use the relations on the sets to solve this problem, and do a check with the VENN diagram.

# Exercise 07

We give the results of the baccalaureate for the following different categories:

| Result⟍Class | Obtained | Failed | Total | % |
|---|---|---|---|---|
| A | | 938 | | 28% |
| B | | 1036 | 2400 | |
| C | | | | 15% |
| D | 1978 | | | |
| Total | | | 9600 | |
| % | 65% | | | |

- *Complete the table*;

A student is asked at random:

- What is the probability that he is a **category B** student who *obtained* his baccalaureate?
- What is the probability that he is a **category B** *baccalaureate graduate*?

## Exercise 08

A survey of **500 students** gave the following statistics:

**186** *students* study *economics*; **295** *students* study *statistics*; **329** *students* study *mathematics*;

**83** *students* study *economics* and *mathematics*; **217** *students* study *mathematics* and *statistics*;

**63** *students* study *economics* and *statistics*;

- How many students study the 03 modules together?

## Exercise 09

A survey of **60 people** revealed that:

**25** *people* read the newspaper **M**; **26** read newspaper **C**; **26** read newspaper **N**;

**11** read newspapers **M** and **C**; **09** read newspapers **M** and **N**; **08** read newspapers **N** and **C**;

**08** people *do not* read any newspaper.

- Represent the VENN diagram;
- How many people read newspapers **M** *and* **C** *and* **N**?
- How many people read *one* newspaper *only*?

## Series N° 02

Events and simple calculations in probability [3-4]

## Exercise 01

In a random experiment, we throw a die twice in succession. We define the following events:

**A**: Number> 4 on the 1st throw;         **B**: Even number on the 1st throw;
**C**: Even number on the 2nd throw;     **D**: Number ≤4 on the 2nd throw;
**E**: Number ≥1 on 1st throw;            **G**: Odd number on the 2nd throw.

- Give the correct relationship (*compatible*; *incompatible*; *independent*; *dependent*) for the following cases:
*A* and *B*  ;  *A* and *C*  ;  *C* and *D*  ;  *E* and *A*  ;  *G* and *B*  ;  *G* and *C*

## Exercise 02

Let **A** and **B** be two events:

If $P(A)=0.8$ ; $P(B)=0.2$ and $P(A\cap B)=0.1$.

- Calculate the value of : $P(A\cup B)$

## Exercise 03

We suppose an event $A$ and its complement $A^c$. Calculate $p(A)$ knowing that:

$$\frac{P(A)}{P(A^c)}=\frac{3}{5}$$

## Exercise 04

Let *A* and *B* be two *incompatible* events. If the probability of *occurrence* of *B* is **three times** ($\times$ 3) that of *A* and the probability of *at least* one of the two is **0.8**.

-    Find the probability that *B* *does not occur*.

**Exercise 05:** We throw a *rigged die* once. The probabilities of *occurrence* of *one of the odd numbers* among these numbers **are equal** (1 or 3 or 5 they have the same probabilities), and the same for the *even numbers*. If we know that the probability of an *even* number is **twice** that of an *odd* number.
-    Calculate the probability of having a **3**;
-    Calculate the probability of having a **4**;
-    Calculate the probability of having a number $\geq$**4**.

## Series N° 03

Combinatorial analysis [1, 2, 5, 6]

## Exercise 01

•   Show that : $n^2 = C_n^2 + C_{n+1}^2$ ;

For what value of n, the following equations are correct:

•
$$0.25 \times A_{n+1}^6 = A_n^6$$

•
$$\frac{7}{6} \times A_n^2 + 7 \times C_n^2 = 4 \times C_{n+1}^3$$

## Exercise 02

In computer science, we use the *binary* system to encode characters. *A bit* (binary digit) is an element that takes the value **0** or the value **1**.

-   With **8** *binary digits* (one byte), how many characters can you encode?

## Exercise 03

In a sports competition involving **18** *athletes*, *Gold*, *Silver*, and *Bronze medals* are awarded.

- How many possible distributions of the medals are there (before competition of course)?

## Exercise 04

Let **A** be the set of *four-digit* numbers, the *first* being *non-zero*.

1. Calculate the number of elements of **A**.

2. Enumerate the elements of **A**:

- Composed of four *distinct* digits;

- Composed of *at least two identical* digits;

- Composed of four *distinct* digits *other than* 5 and 7.

## Exercise 05

A sports tournament has **8** *teams* involved. Each team must *meet* all the others *only once*.

- How many *matches* should we organize?

## Exercise 06

| 1 | 2 | 3 |
| 4 | 5 | 6 |
| A | B | C |

A **9-key** *keypad* allows you to dial the entry code of a building, using a *letter followed by a number of 3 distinct digits or not*.

1. How many *different codes* can we train?

2. How many *codes* are there *without the number* **1**?

3. How many *codes* have the *number* **1** *at least once*?

4. How many *codes* are there with *distinct digits*?

5.   How many *codes* are there with *at least two identical digits*?

## Exercise 07

In a class of **32** *students*, there are **19 boys** and **13 girls**. We must *elect two delegates*

-   What is the number of *possible choices?*

-   What is the number of choices if it is imposed on a **boy** and a *girl?*

-   What is the number of choices if it is imposed on **2** *boys*?

## Exercise 08

1.   Number the *anagrams* of the word PATRICE.

2.  In each of the following cases, count the *anagrams* of the word PATRICE:

-   *Beginning* and *ending* with a *consonant*;

-   *Beginning* and *ending* with a *vowel*;

-   *Beginning* with a *consonant* and *ending* with a *vowel*;

-   *Beginning* with a *vowel* and *ending* with a *consonant*.

## Exercise 09

We constitute a *group* of **06** *people* chosen *among* **25** *women* and **32** *men,*

1.   How many ways can this group of **6** *people* be *formed*?

2. In each of the following cases, in how many ways can this group be formed with?

-   *Only men*;

-   People of the *same sex*;

-   *At least one woman* and at *least one man.*

## Exercise 10

A *coat rack* has **05** *aligned hooks*. How many separate provisions (without putting two coats on each other)

- For **03** *coats* on these **05** *hooks*?

- For **05** *coats*?

- For **06** *coats*?

## Series N° 04

Probability calculations [1, 2, 5]

## Exercise 01

- A statistic done on *2nd year Mechanical Engineering students* showed that:
- **15%** of the students are *excellent*; among them **25%** are *girls*;
- **38%** are of *average level*, where **68%** are *boys*;
- The **remaining** students are of *low level* and the number of *girls* is *equal* to that of *boys*.
1. Carefully trace the *probability tree diagram*, including all the necessary details;

A student is asked at random:

2. What is the probability that the student is a *girl* in the *third category* (the weak).

## Exercise 02

The realization of a *construction project* requires a series of works: *excavation* (**E**), *foundation* (**F**), and *structure* (**S**). The *probabilities* that they will be *completed according to the schedule* are **0.7**, **0.8** and **0.9** *respectively*. It is assumed that the events are *mutually independent*. Calculate the probabilities of the following events:

3. **K**: *Project completed* according to schedule;
4. **G**: *Excavation completed* and *at least one of the other two stages completed* in time;
5. **H**: *Only one* of the three stages was *completed* in time.

## Exercise 03

A transmission shaft manufacturing company has **three** *production lines*, which manufacture the *same type of shaft*. **40%** of *the shafts* come out of the *first chain*, while **35%** come out of the *second* and **the rest** of the *third*. **Failure tests** have shown that **2%** of the *shafts* coming out of the *first chain* have defects. The *second* and *third chain*s have defects of **4%** and **5%,** *respectively*.

1. We take a shaft at random. What is the probability that it is *not faulty*?
2. If it is *defective* (faulty), what is the probability that it will be *made by chain three*?

## Exercise 04

An urn contains **15** *balls* identical to the touch, of which **07** are *white* and the **remainder** *black*. We draw one ball; we note its color and then we put it back into the urn before making another draw. For transparency, the urn is shaken before each draw. We made a total of **05** draws.

If we inform you that we are interested in obtaining the white ball:

- What is the probability of having *exactly* **one** *white ball* after the 05 draws?
- What is the probability of obtaining a *maximum of* **02** *white balls* after the 05 draws?
- Carefully draw the *bar graph* of the experiment;
- Calculate the *mathematical expectation*.

## Series N° 05

Random variable [2, 7]

## Exercise 01

A random variable $x$ is given by the following probability law:

| $x_i$ | -2 | -1 | 0 | 1 | 2 | 3 |
|---|---|---|---|---|---|---|
| $p(x=x_i)$ | 0.3 | 0.05 | 0.1 | 0.05 | 0.2 | $p$ |

Let $F(x)$ be its distribution function.

1. Calculate $p$ ;
2. Calculate $F(0.5)$ ;
3. Calculate $E(x)$ ;
4. Calculate $\sigma(x)$.

## Exercise 02

A *non-rigged* die is made up of **six** *colored sides*: **Three** *red*, **Two** *green*, and **the Last** *yellow* (**3R**; **2G**; 1Y). A *number of points* are *assigned* to each color: *Red* = **0** pt; *Green* = **01** pt; *Yellow* = **02** pts.

We *throw* the dice **twice** *in a row* and we record the **sum** *of the points* obtained which represents our *random variable x* for this experiment.

1. Give the law of probability for this x;
2. Calculate: $E(X)$; $V(X)$ et $\sigma(X)$.

## Exercise 03

**Sixteen** (**16**) *passengers **bought*** their travel tickets at station **A**.

- **Seven** (**07**) *go* to *station* **B** (The price of the trip is **50 AD** *per passenger*);
- **Five** (**05**) *go* to *station* **C** (The price of the trip is **60 AD** *per passenger*);
- **The rest** *go* to *station* **D** (The price of the trip is **75 AD** *per passenger*).
1. **One** (**01**) *passenger* is *chosen at random*. Let $x$ be the random *variable* that *assigns* the **price of his ticket** to the ***passenger***.

a. Define the probability law of $x$;

b. Calculate the mathematical expectation of $x$;

1. **Three** (**03***) passengers* are *chosen at random*.

a. Calculate the probability that the *destinations* (direction of travel) of the passengers are *different*;
b. Calculate the probability that *at least* **one** (**01**) *passenger* is heading to *station* **B**;
c. If you are informed that **03** *passengers* are heading to the *same station*. What is the probability that the **three** (**03**) passengers are heading *towards* **B**?

## Series N° 06

Distribution laws [4-6]

### Exercise 01

We consider a random variable $X$ which follows a *binomial distribution* of *parameters* **20** and **0.4**.

1. Calculate : $p(X=3)$ ; $p(X=17)$ ; $p(X=10)$.
2. Calculate: $p(X\leq1)$ ; $p(X\geq18)$ ; $p(X\leq15)$ ; $p(X\geq10)$.

### Exercise 02

In an *oil-rich region*, the probability that a *drilling will lead* to an *oil slick* is **0.1**.

1. Justify that the realization of a borehole can be assimilated to a Bernoulli test;
2. We perform 09 boreholes:
a. What *hypothesis* must be formulated so that the random variable *x* corresponding to the number of boreholes that led to an oil slick follows a binomial law?
b. Under this assumption, calculate the probability that *at least one borehole* leads to an *oil slick*.

### Exercise 03

A component manufacturer produces electronic resistors. The probability that a resistor is defective is equal to $5\times10^{-3}$.

In a batch of **1000** *resistors*, what is the probability of having:

a. *Exactly two* defective resistors?
b. *At most two* defective resistors?
c. *At least two* defective resistors?

### Exercise 04

If **one** in a *hundred people* in a population is a ***centenarian*** (their age reaches 100), what is the probability of finding *at least* **one** centenarian:

a. *Among* **100** *people* chosen at random?

b.  *Among* **200** *people?*

## Exercise 05

A machine makes steel washers. The diameter of a washer follows a normal law with mean $\mu = 90mm$ and standard deviation $\sigma = 0.16mm$.

a.  What is the probability that the *diameter* of a random washer is *outside* the interval $[89.7mm; 90.3mm]$?;
b.  Find the *number d* such that the proportion of washers with a diameter between **90-d** and **90+d** is **90%**.

**1.**We reject the washers whose diameter is *outside* the interval $[89.7mm; 90.3mm]$. The *probability* that a washer will be judged to be *defective* is **0.06**. From a lot containing a very large number of washers, $N$ pieces are drawn. We call $X$ the random variable which -to this test- associates the *number of defective washers*.

a.       We draw **04** *pieces* (i.e. $N=4$):

- What is the law of probability that $X$ follows? What are its parameters?
- What is the expression of $p(X=k)$ ?
- Calculate the probability $p_1$ so that we have *exactly* **01** acceptable pieces;
- Calculate the probability $p_2$ so that we have *at least* **02** acceptable pieces.

b.       We draw **50** *pieces*: We admit that the law followed by $X$ can be approximated by a ***Poisson law***, *the parameter of which will be specified.*

- What then is the expression of $p(X=k)$?;
- Calculate the probability $p_3$ of having *no defective* (**0**) *pieces*;

Calculate the probability $p_4$ of having *at most* **02** *defective pieces.*

# Series of Statistics Exercises

**Series N° 7**

**STATISTICS**

**Exercise 01**

Determine for the series below, the *mean*, the *mode* and the *median* [8, 9, 12, 14].

**Series 1:**          1    3    6    8    9    11    14    17    20    24    29

**Series 2:**

| $x_i$ | 02 | 04 | 07 | 14.5 | 10.0 | 8.8 | 7.0 | 9.0 | 3.0 |
|-------|----|----|----|------|------|-----|-----|-----|-----|
| $n_i$ | 02 | 06 | 01 | 01   | 07   | 05  | 02  | 04  | 03  |

**Exercise 02**

Statistics made on **40** *employers* of a company about their *late arrivals* (in minutes) *of morning* entry are given in the table below:

| Class | [A-B] | [B-C] | [C-D] | [D-E] |
|-------|-------|-------|-------|-------|
| **Center of the class** | 7.5 | 12.5 | 17.5 | 22.5 |
| **Workforce (Frequency)** | 15 | 10 | 06 | 09 |

1. Determine the *population*, the *random variable* and give *its type*;
2. If you are informed that the *lengths of the classes are equal*. Calculate the *length* of the class and define *the limits* of each class;
3. Calculate the *mean, standard deviation* and *coefficient of variation* and give your opinion on the dispersion of the series ;
4. Carefully draw the *histogram of the series* and mention the *corresponding polygon* on it.

## Exercise 03

A quality control inspector extracted from his database, *a sample* of **40** *weeks* where he noted *X*, the *number of work accidents recorded per week*. He obtained the following results:

2  0  4  2  2  1  3  2  0  5  4  3  2  4  5  6  6  4  2  0

3  4  4  2  6  2  4  3  0  4  3  4  3  3  5  5  4  2  2  1

- Complete the following table and draw the representative diagram.

| Nbr of accidents per week | Frequency | Relative frequency | Cumulative relative frequency |
|:---:|:---:|:---:|:---:|
| . | . | . | . |
| . | . | . | . |
| . | . | . | . |
| . | . | . | . |
| **Total** | | | |

## Exercise 04

Let *X* be the *daily receipts* (*in dollars*) of a *small store*. A sample of size *n*=**40** *days* was randomly selected which yielded the following results:

16.00   58.50   68.20   78.00   79.45   142.20   145.3   186.70   209.05   216.75

219.70  247.75  249.10  256.00  257.15  262.35  268.60  269.60  270.15  284.45

319.00  332.00  343.29  350.75  354.90  372.60  383.20  389.20  404.55  420.20

428.50  432.40  444.60  446.80  456.10  458.10  493.95  511.95  521.05  621.35

The number of classes to be formed is: $K = 1 + \dfrac{10}{3}\log(n)$

The amplitude of each class equal to: $A = \dfrac{X_{max} - X_{min}}{K}$ .

**Note:** The results for $K$ and $A$ must be rounded off with excess.

| X (Recipe/Day) | Frequency | Relative frequency | Cumulative relative frequency |
|---|---|---|---|
| [ ... , ... ] | - | - | - |
| . | . | . | . |
| . | . | . | . |
| . | . | . | . |
| . | . | . | . |
| Total | | | |

- Complete the following table and draw the representative diagram.

## Exercise 05

Here are the distribution frequencies of a population.

Calculate the following:

- The *arithmetic mean*;
- The *mode* with explanation of the result;
- The *median* with an explanation of the result;
- The *first quartile* with an explanation of the result.

| $x_i$ | [1-1.5] | [1.5-2] | [2-2.5] | [2.5-3] | [3-3.5] | [3.5-4] |
|---|---|---|---|---|---|---|
| $n_i$ | 13 | 40 | 96 | 65 | 30 | 50 |

# Series Correction Probabilities

## SERIES 1

**Sets**

### Exercice 01

- $I\backslash J=\{2\;;\;5\;;\;8\}\;\Rightarrow\;x=\{2\;;\;5\;;\;8\}$
- $I=\{k\;;\;b\;;\;x\;;\;a\;;\;c\}$ and $J=\{b\;;\;d\;;\;y\;;\;h\;;\;a\}\Rightarrow\{k\;;\;x\;;\;c\}=\{d;\;y\;;c\}\Rightarrow y=\{k\;;\;c\}$ and $x=\{d;\;c\}$

This solution is *not unique*, we can add on both sides, one element or more of our choice, but the same element (s).

- $I=\{19\;;\;x\;;\;y\;\}$ and $J=\{24\;;\;20\;;\;7\;;\;25\;;\;19\}\Rightarrow\{x\;;\;y\}=\{24\;;\;20\;;\;7\;;\;25\}$

In this case also we have infinity of possibilities.

### Exercice 02

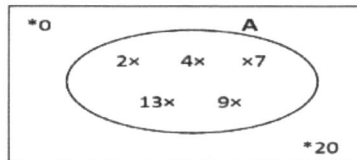

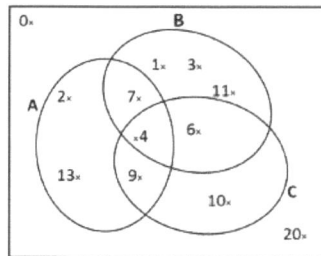

### Exercice 03

## Exercice 04

- We define $B$ and $C$ in extension (i. e. enumeration)

$B=\{5;6;7;8;9\}$   ;   $C=\{0;2;4;6;8\}$

So : $B\cap C=\{6;8\}$

- $A\cup(B\cap C)=\{1;2;3;4;5;7\}\cup\{6;8\}=\{1;2;3;4;5;6;7;8\}$
- $(A\cup B)\cap(A\cup C)=\{1;2;3;4;5;6;7;8\ ;9\}\cap\{0;1;2;3;4;5;6;7;8\}=\{1;2;3;4;5;6;7;8\}$
  So :  $A\cup(B\cap C)=(A\cup B)\cap(A\cup C)$
- $A^C=S\backslash A=\{0;6;8;9\}$
- $C^C=S\backslash C=\{1;3;5;7;9\}$
- $A^C\cap C^C=\{9\}$
- $(A\cup C)^C=S\backslash(A\cup C)=S\backslash\{0;1;2;3;4;5;6;7;8\}=\{9\}$
  So :  $(A\cup C)^C = A^C\cap C^C$               (**1ˢᵗ** of De-Morgan)
- $A^C\cup C^C=\{0;6;8;9\}\cup\{1;3;5;7;9\}=\{0;1;3;5;6;7;8;9\}$
- $(A\cap C)^C=S\backslash(A\cap C)=S\backslash\{2;4\}=\{0;1;3;5;6;7;8;9\}$

So :  $(A\cap C)^C = A^C\cup C^C$               (**2ⁿᵈ** of De-Morgan)

## Exercise 05

We draw the VENN diagram for the two sets. *T* for those who practice *Tennis* and *M* for those who practice *Swimming*.

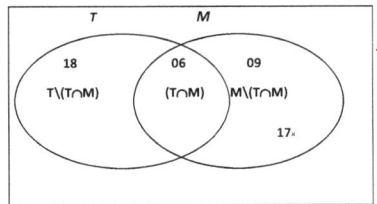

- We have the total set of people questioned Card($S$)=50

So, those who do not play any sport are: $S\backslash(T\cup M)$ and their number=50-(18+6+9)=17.

- The *probability* of having a person playing **only one** sport is equal:

$$p = \frac{Card\left(T \setminus (T \cap M)\right) + Card\left(M \setminus (T \cap M)\right)}{Card(S)} = \frac{18 + 9}{50} = 0.54$$

- The *probability* of having a person playing **Tennis** is equal:

$$p = \frac{Card\left(T \setminus (T \cap M)\right) + Card\left(T \cap M\right)}{Card(S)} = \frac{18 + 6}{50} = 0.48$$

## Exercise 06

As done for the previous exercise, we name those who owned a *European car* by *E* and those, an *Asian car* by *A*.

So, we have: $Card(E) = 75$   ;    $Card(A) = 60$   ;    $Card(J \cap A) = 45$

- The number of people who did not own a car (neither *European* nor *Asian*) is in total *minus* the number of those who have ever owned a car or two.

$$Card(NC) = Card(S) - Card(E \cup A) = Card(S) - \left[Card(E) + Card(A) - Card(E \cap A)\right]$$

$$= 100 - (75 + 60 - 45) = 10$$

- The number of people who have owned a *European* car but not an *Asian* car is defined by: $Card(E \setminus (E \cap A))$

So: $Card(E \setminus (E \cap A)) = Card(E) - Card(E \cap A) = 75 - 45 = 30$

- Verification with the VENN diagram:

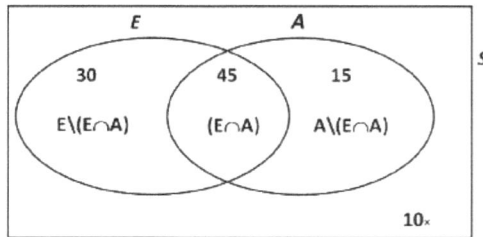

## Exercise 07

The table can be completed in several ways. After all the calculations, we arrive at the following values in the table:

| Result / Class | Obtained | Failed | Total | % |
|---|---|---|---|---|
| A | 1750 | 938 | 2688 | 28% |
| B | 1364 | 1036 | 2400 | 25% |
| C | 1148 | 292 | 1440 | 15% |
| D | 1978 | 1094 | 3072 | 32% |
| Total | 6240 | 3360 | 9600 | 100% |
| % | 65% | 35% | 100% | |

- The probability that he is a class *B* student, who obtained his baccalaureate:
$$p = \frac{1364}{9600} = 14.2\%$$
- The probability that he is a category *B* baccalaureate graduate:
$$p = \frac{1364}{6240} = 21.85\%$$

(It is a case of *conditional probability* that we will see later)

# Exercise 08

- We call *E* the set of students studying *Economy*;

- We call *M* the set of students studying *Math*;

- We call *S* the set of students studying *Statistics*;

- We call *EM* the set of students studying *Economy* and *Math*;

- We call *ES* the set of students studying *Economy* and *Statistics*;

- We call *MS* the set of students studying *Math* and *Statistics*;

-   We call **EMS** the set of students studying *Economy, Math* and *Statistics* simultaneously.

From the exercise' details, all the cardinals of the sets mentioned above are known except for the last set. To have it, we use the law seen during the course on sets (at the end of the chapter):

$$Card(E \cup M \cup S) = Card(E) + Card(M) + Card(S) - Card(EM) - Card(ES) - Card(MS)$$
$$+ Card(EMS)$$

$$500 = 186 + 329 + 295 - 217 - 63 - 83 + Card(EMS) \quad \Rightarrow \quad Card(EMS) = 53$$

Using this result and the data, we arrive at the following VENN diagram:

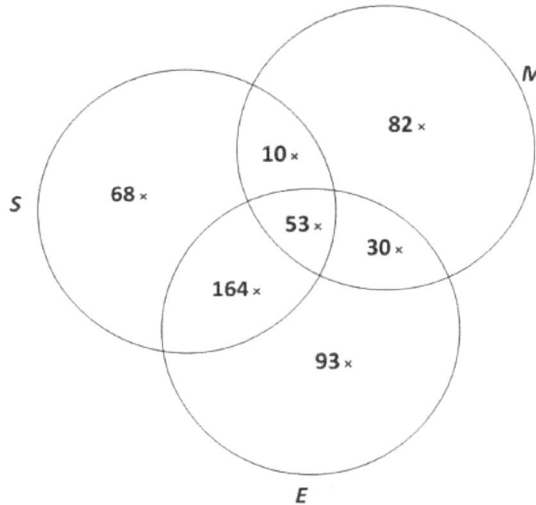

**Series 2:**  Events and simple calculations in probability

**Exercise 01**

We have (*L: Launch*):

$$A = \{5;6\}_{1^{st} L} \quad ; \quad B = \{2;4;6\}_{1^{st} L} \quad ; \quad C = \{2;4;6\}_{2^{nd} L}$$

$$D = \{1;2;4;6\}_{2^{nd} L} \quad ; \quad E = \{1;2;3;4;5;6\}_{1^{st} L} \quad ; \quad G = \{1;3;5\}_{2^{nd} L}$$

So, we have the following relationships between the events

$A$ and $B$ $\longrightarrow$ Compatible    $E$ and $A$ $\longrightarrow$ Compatible

$A$ and $C$ $\longrightarrow$ Independent    $G$ and $B$ $\longrightarrow$ Independent

$C$ and $D$ $\longrightarrow$ Compatible    $G$ and $C$ $\longrightarrow$ Incompatible

## Exercise 02

We have:    $p(A \cup B) = p(A) + p(B) - p(A \cap B)$

$$\Rightarrow p(A \cup B) = 0.8 + 0.2 - 0.1 = \boxed{0.9}$$

## Exercise 03

We have:
$$p(A^C) = 1 - p(A)$$

And we have:
$$\frac{p(A)}{p(A^C)} = \frac{3}{5}$$

$$\Rightarrow \frac{p(A)}{1 - p(A)} = \frac{3}{5} \Rightarrow 5 \times p(A) = 3 - 3 \times p(A) \Rightarrow 8 \times p(A) = 3 \Rightarrow \boxed{p(A) = \frac{3}{8}}$$

## Exercise 04

We have:
$$A \text{ and } B \text{ incompatibe } \Rightarrow p(A \cap B) = 0$$

$$\Rightarrow p(A \cup B) = p(A) + p(B) = 0.8$$

And we have:   $p(B) = 3 \times p(A)$

$$\Rightarrow p(B) = 3 \times [0.8 - p(B)] \Rightarrow 4 \times p(B) = 2.4 \Rightarrow p(B) = 0.6 \Rightarrow \boxed{p(B^C) = 0.4}$$

## Exercise 05

We have:
$$O: for\,Odd \qquad E: for\,Even$$

We have:
$$p(E) = 2 \times p(O)$$

We have:
$$O \text{ and } E \text{ incompatibe}$$

So:

$$3 \times p(O) + 3 \times p(E) = 1 \quad (all\ issues)$$

$$\Rightarrow\ 3 \times p(O) + 6 \times p(O) = 1$$

$$\Rightarrow\ 9 \times p(O) = 1\ \Rightarrow\ \boxed{p(O) = 1/9}$$

- Determination of $p(3)$, $p(4)$ and $p(\geq 4)$ :

$$p(O) = p(1) = p(3) = p(5) = \boxed{1/9}$$

$$p(E) = p(2) = p(4) = p(6) = \boxed{2/9}$$

$$p(\geq 4) = p(4) + p(5) + p(6) = \boxed{5/9}$$

**Series 3:**  Combinatorial analysis

Answer for the values of **n** ensuring the equations:

- $C_n^2 + C_{n+1}^2 = \dfrac{n!}{2! \times (n-2)!} + \dfrac{(n+1)!}{2! \times (n+1-2)!} = \dfrac{1}{2}\left[n \times (n-1) + (n+1) \times n\right]\dfrac{(n-2)!}{(n-2)!} = \dfrac{1}{2}\left[n^2 - n + n^2 + n\right] = n^2$

- $0.25 \times A_{n+1}^6 = A_n^6$

$$\Rightarrow 0.25 \times \frac{(n+1)!}{((n+1)-6)!} = \frac{n!}{(n-6)!} \Rightarrow \frac{1}{4} \times \frac{(n+1)!}{(n-5)!} = \frac{n!}{(n-6)!} \Rightarrow \frac{(n+1)!}{(n-5)!} = 4 \times \frac{n!}{(n-6)!}$$

$$\frac{(n+1)!}{(n-5)!} = \frac{(n+1) \times n!}{(n-5) \times (n-6)!} = \frac{n+1}{n-5}\frac{n!}{(n-6)!}$$

$$\Rightarrow \frac{n+1}{n-5} = 4 \Rightarrow n+1 = 4 \times (n-5) \Rightarrow n = 7$$

$$\frac{7}{6} \times A_n^2 + 7 \times C_n^2 = 4 \times C_{n+1}^3$$

-

$$\Rightarrow 7 \times A_n^2 + 42 \times C_n^2 = 24 \times C_{n+1}^3 \Rightarrow 7 \times \frac{n!}{(n-2)!} + 42 \times \frac{n!}{2! \times (n-2)!} = 24 \times \frac{(n+1)!}{3! \times (n-2)!}$$

$$\Rightarrow 7 \times \frac{n!}{(n-2)!} + \frac{42}{2!} \times \frac{n!}{(n-2)!} = \frac{24}{3!}(n+1) \times \frac{n!}{(n-2)!}$$

$$\Rightarrow 7 + 21 = 4 \times (n+1) \Rightarrow n = 6$$

## Exercise 01

Examples for an Octet:

$$
\begin{array}{cccccccc}
0 & 0 & 0 & 0 & 0 & 0 & 0 & 0 \\
0 & 0 & 0 & 0 & 0 & 0 & 0 & 1 \\
1 & 1 & 0 & 1 & 0 & 0 & 0 & 1 \\
\vdots & \vdots & \vdots & \vdots & \vdots & \vdots & \vdots & \vdots \\
1 & 1 & 1 & 1 & 1 & 1 & 1 & 1 \\
\end{array}
$$

So, each cell can be chosen in two ways **0** or **1**. The total number will be according to the principle of multiplication:

$$N = \prod_{i=1}^{8} n_i = 2 \times 2 \times 2 \times 2 \times 2 \times 2 \times 2 \times 2 = 2^8 = \boxed{256}$$

## Exercise 02

With the precision of one *type of metal* for a *medal*, the order becomes important. In addition, only **03** will have medals *among* the **18** *athletes*. Therefore:

$$N = A_n^k = \frac{18!}{(18-3)!} = \frac{18!}{15!} = 18 \times 17 \times 16 = \boxed{4896}$$

## Exercise 03

- *A* is the set of ***four**-digit* numbers (*the first* on the *left not equal* to **0**).

So: the *first* on the left has **09** possibilities of choice, the *second* **10**, the *third* **10** and the *last* also **10**.

$$N = \prod_{i=1}^{4} n_i = 9 \times 10 \times 10 \times 10 = \boxed{9000}$$

- For **four different** digits: $N = A_9^1 \times A_9^3 = 9 \times \dfrac{9!}{(9-3)!} = 9 \times 9 \times 8 \times 7 = \boxed{4536}$

- For *at least two identical* digits: *i.e.*, two or three or four. This is exactly the opposite event of four different digits. Therefore :

$$N = 9000 - 4536 = \boxed{4464}$$

- For *four different* digits other than **5** and/or **7**:

$$N = A_7^1 \times A_7^3 = 7 \times \dfrac{7}{(7-3)!} = 7 \times 7 \times 6 \times 5 = \boxed{1470}$$

## Exercise 04

We have **08** *teams* and *each team will play* with all the remaining seven a *no-return game*. So, we can say that team *A* is playing with *B* or vice versa. *I.e.* the order *A-B* or *B-A* does not matter (if there has been a *return match*, the *order becomes important*). In addition, for each match, we take *two teams* from the **08**. So, it is the case of a *combination without repetition* (the teams are different and the team cannot play against itself).

$$N = C_8^2 = \dfrac{8!}{2! \times (8-2)!} = \dfrac{8!}{2! \times 6!} = \dfrac{8 \times 7}{2} = \boxed{28}$$

## Exercise 05

The code has the form:  **Letter - Number - Number - Number**

We have 03 letters: *A*, *B* and *C*          ///   We have 06 numbers: **1**, **2**, **3**, **4**, **5** and **6**

- $N = \prod_{i=1}^{4} n_i = 3 \times 6 \times 6 \times 6 = \boxed{648}$  (the numbers *can be repeated*)

- $N = \prod_{i=1}^{4} n_i = 3 \times 5 \times 5 \times 5 = \boxed{375}$  (the digits can *be repeated without* the possibility of *choosing* the **1**)

- A code with the existence of the number **1** at *least once*. *i.e.* the opposite event to that of the previous question:

$$N_{nub1} = N_{no\,cond} - N_{without\,nub1} = 648 - 375 = \boxed{273}$$

- A code with *distinct* (different) digits :

$$N = \prod_{i=1}^{4} n_i = 3 \times 6 \times 5 \times 4 = \boxed{360}$$

$$or \quad N = A_3^1 \times A_6^3 = 3 \times \frac{6!}{3!} = 3 \times 6 \times 5 \times 4 = \boxed{360}$$

- A code with *at least two identical digits. i.e.* the opposite event of the previous one:

$$N_{at\,least\,tow\,dig} = N_{no\,cond} - N_{diff\,dig} = 648 - 360 = \boxed{288}$$

## Exercise 06

- The number of possible choices is consistent with a combination without repetition, as no task was specified for the chosen students. So, the order is not important. In addition, we took only two out of **32** with the impossibility of repeating a student:

$$N = C_{32}^2 = \frac{32!}{2! \times (32-2)!} = \frac{32 \times 31}{2} = \boxed{496}$$

- For a boy and a girl: $N = A_{19}^1 \times A_{13}^1 = \frac{19!}{18!} \times \frac{13!}{12!} = 19 \times 13 = \boxed{247}$

*Note*: For choosing a single item (student here), we can use the combination as the arrangement ($k=1$).

- For two boys (order not important): $N = C_{19}^2 = \frac{19!}{2! \times 17!} = 19 \times 9 = \boxed{171}$

## Exercise 07

- *The anagrams of the word* PATRICE *are the case of a permutation without repetition (we took all the letters, so **k** equals **n** and the letters are different):*

$$P_n = n! = 7! = \boxed{5040}$$

- *An anagram that begins and ends with a consonant*: We choose **02** consonants among the **04** that we have (P, T, R, and C), then the **05** remaining letters to complete the word will be chosen from the remaining **05** letters.

$$N = A_4^2 \times P_5 = \frac{4!}{2!} \times 5! = 12 \times 120 = \boxed{1440}$$

- *Starts and ends with a vowel*: $N = A_3^2 \times P_5 = \frac{3!}{1!} \times 5! = 6 \times 120 = \boxed{720}$

- *Starts with a consonant and ends with a vowel*:

$$N = A_4^1 \times P_5 \times A_3^1 = 4 \times 5! \times 3 = 4 \times 120 \times 3 = \boxed{1440}$$

- *Starts with a vowel and ends with a consonant*:

$$N = A_3^1 \times P_5 \times A_4^1 = 3 \times 5! \times 4 = 3 \times 120 \times 4 = \boxed{1440}$$

## Exercise 08

1. We want to form groups of **06** people from **57** people, without any precision. So, the order doesn't matter and repetition is not possible. This, is the case for a combination without repeating:

$$N = C_{57}^6 = \frac{57!}{6! \times 51!} = \boxed{36288252}$$

a. *Groups of men only*: $N = C_{32}^6 = \frac{32!}{6! \times 26!} = \boxed{906192}$

b. *People of the same sex* (group of **06** women or **06** men):

$$N = C_{32}^6 + C_{25}^6 = \frac{32!}{6! \times 26!} + \frac{25!}{6! \times 19!} = \boxed{1083292}$$

a. For groups containing *at least* one woman and *at least* one man, we have **05** situations to account:

$(1w+1m+4m)+(1w+1m+3m+1m)+(1w+1m+2m+2w)+(1w+1m+1m+3w)+(1w+1m+4w)$

$$N = C_{25}^1 \times C_{32}^1 \times C_{32}^4 + C_{25}^1 \times C_{32}^1 \times C_{32}^3 \times C_{25}^1 + C_{25}^1 \times C_{32}^1 \times C_{32}^2 \times C_{25}^2$$

$$+ C_{25}^1 \times C_{32}^1 \times C_{32}^1 \times C_{25}^3 + C_{25}^1 \times C_{32}^1 \times C_{25}^4 = \boxed{305204960}$$

We can see that these **05** situations are *the complements* of the **02** cases where the group is formed of ***women*** or ***men*** *only*. So, we can calculate directly as follows :

$$N = C_{57}^6 - \left[ C_{32}^6 + C_{25}^6 \right] = \boxed{305204960}$$

## Exercise 09

We have **05** places on the coat racks.

a.  For *three coats*, we *have enough places* to place them. This is the case of an arrangement of 3 among 5 without repetition, because each coat takes an independent place, and the coats are different.

$$N = A_5^3 = \boxed{60}$$

*Note:* The order here is important. We say *for example*: *coat* **1** in place 3, *coat* **2** in place 1 and *coat* **3** in place 4 ... etc.

b.   For **05** *coats*, we have exactly **05** *places to occupy*. This is the case of a permutation without repetition:

$$P_n = 5! = \boxed{120}$$

c.  For **06** *coats*, we have **0** *possibilities* because each place is occupied by a single coat and therefore, the **6**[th] *coat* must be placed with another on the same place.

*If we accept* to place **02** *coats* on *one place*, we will have *permutations* for the **05** *first coats* and **05** *possibilities* for the **6**[th] *coat*.

$$N = 5 \times P_5 = 5 \times 5! = \boxed{600}$$                    (02 maximum coats after filling all the places)

**Series 4:**   Probability calculations

**Exercise 01**

1.Probability tree diagram with details: A student is questioned at random:

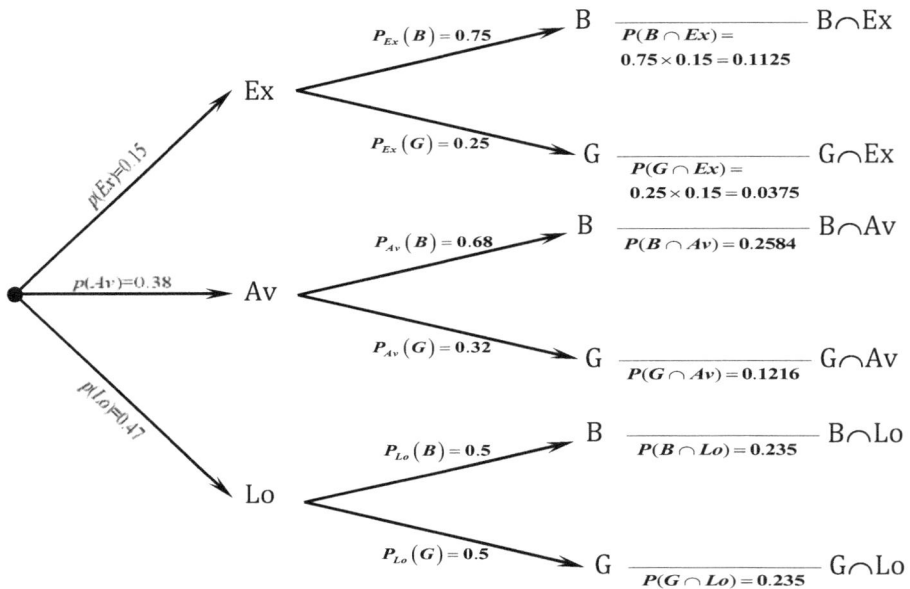

-On the probability tree, we move from left to right. If in the question we specify the opposite path (as here), this implies that we want to know the origin of the event sought. Such a situation is none other than the application of *Bayes' theorem*. So, we are looking for: $P_G(Lo)$.

We have:   $P_G(Lo) = \dfrac{P(Lo) \times P_{Fa}(G)}{P(G)} = \dfrac{0.47 \times 0.5}{0.15 \times 0.25 + 0.38 \times 0.32 + 0.47 \times 0.5} \simeq 0.5963$

*i.e.* that almost 60% of girls are of low level.

**Exercise 02**
The project is carried out on three successive stages: $E$ then $F$ then $S$.

So, If one of the steps is completed within its estimated time, we say that this event (*estimated time*) is completed. Otherwise, we say the opposite event (*delay*) is achieved.
Therefore, the probability tree of the problem is as follows (below):

- The event **K** *is realized*: $P(K) = P(EFS) = 0.7 \times 0.8 \times 0.9 = 0.504$
- The        event        **G**        is        realized:
  $P(G) = P(EFS^c) + P(EF^cS) = 0.7 \times 0.8 \times 0.1 + 0.7 \times 0.2 \times 0.9 = 0.182$
- The event **H** is realized: $P(H) = P(EF^cS^c) + P(E^cFS^c) + P(E^cF^cS) =$

$$0.7 \times 0.2 \times 0.1 + 0.3 \times 0.8 \times 0.1 + 0.3 \times 0.2 \times 0.9 = 0.092$$

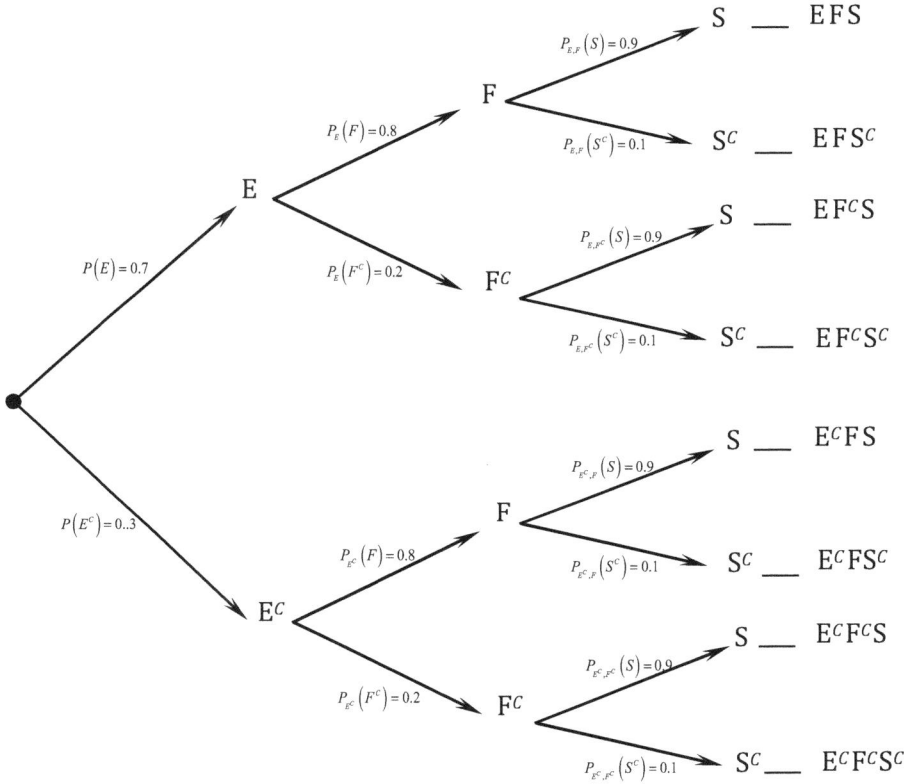

## Series 5: Random variable

## Exercise 01

- The sum of the probabilities of $x_i$ must absolutely *equal* **1**. Thus:

$$p = 1 - (0.3 + 0.05 + 0.1 + 0.05 + 0.2) = 0.3$$

- The value of the distribution function **F** at a value of **x**, equals the sum of the probabilities of all the $x_i$ located before and exactly to the fixed $x_i$. Therefore:

$$F(0.5) = p(x \le 0.5) = 0.3 + 0.05 + 0.1 = 0.45$$

- $$E(x) = \sum_{i=1}^{6} p_i \times x_i = 0.3 \times (-2) + 0.05 \times (-1) + 0.1 \times (0) + 0.05 \times (1) + 0.2 \times (2) + 0.3 \times (3) = 0.7$$

$$V(x) = \sum_{i=1}^{6} p_i \times x_i^2 - E(x)^2$$

$$= 0.3 \times (-2)^2 + 0.05 \times (-1)^2 + 0.1 \times (0)^2 + 0.05 \times (1)^2 + 0.2 \times (2)^2 + 0.3 \times (3)^2 - (0.7)^2 = 4.31$$

- $$\sigma(x) = \sqrt{V(x)} \simeq 2.077$$

## Exercise 02

The diagram below summarizes the random experiment and the calculation of probabilities, in addition to the values of the random variable $X$.

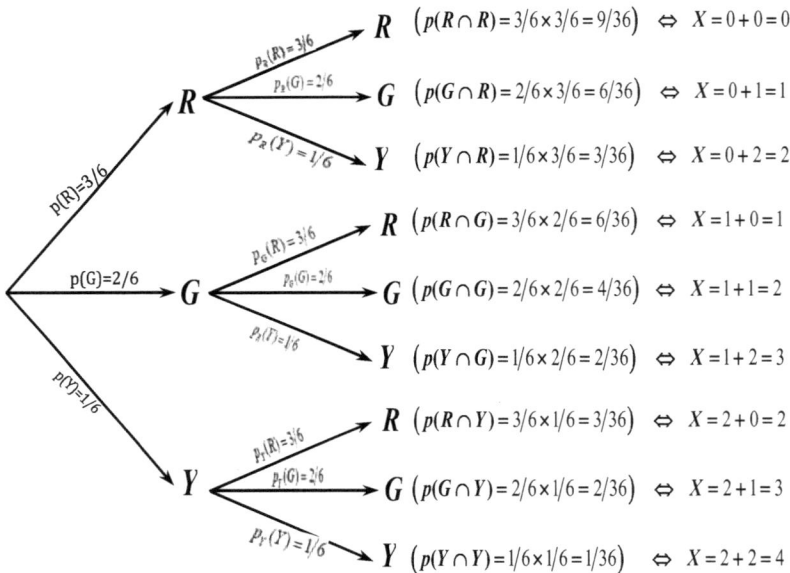

So, the probability law of $X$ is:

| $X$ | 0 | 1 | 2 | 3 | 4 |
|-----|-----|-----|-----|-----|-----|
| $p(X=x_i)$ | 9/36 | 12/36 | 10/36 | 4/36 | 1/36 |

*Calculation of: $E(X)$ ; $V(X)$ and $\sigma(X)$ :*

$$E(X) = \sum_{i=1}^{5} p_i \times x_i = \frac{9}{36} \times 0 + \frac{12}{36} \times 1 + \frac{10}{36} \times 2 + \frac{4}{36} \times 3 + \frac{1}{36} \times 4 = \frac{48}{36}$$

$$E(X) \simeq 1.334$$

$$V(X) = \sum_{i=1}^{5} p_i \times (x_i - E(X))^2$$

$$= \frac{9}{36} \times \left(0 - \frac{48}{36}\right)^2 + \frac{12}{36} \times \left(1 - \frac{48}{36}\right)^2 + \frac{10}{36} \times \left(2 - \frac{48}{36}\right)^2 + \frac{4}{36} \times \left(3 - \frac{48}{36}\right)^2 + \frac{1}{36} \times \left(4 - \frac{48}{36}\right)^2$$

$$= \frac{1}{36^3} \left[9 \times 48^2 + 12 \times 12^2 + 10 \times 24^2 + 4 \times 60^2 + 1 \times 96^2\right] = \frac{51840}{36^3}$$

$$V(X) \simeq 1.111$$

$$\sigma(X) = \sqrt{V(X)} = \sqrt{\frac{51840}{36^3}}$$

$$\sigma(X) \simeq 1.054$$

## Exercise 03

1.

a. **X** represents the price of the ticket for each passenger. We therefore have 03 values for

**X**: 50*AD*; 60*AD* or 70*AD*.

The probabilities of the different $x_i$:

$$p(X = 50DA) = \frac{7}{16} \quad ; \quad p(X = 60DA) = \frac{5}{16} \quad ; \quad p(X = 75DA) = \frac{4}{16}$$

So, the probability law of this *RV* is:

| X (DA) | 50 | 60 | 75 |
|---|---|---|---|
| p(X=$x_i$) | 7/16 | 5/16 | 4/16 |

b.
$$E(X) = \sum_{i=1}^{3} p_i.x_i = \frac{7}{16} \times 50 + \frac{5}{16} \times 60 + \frac{4}{16} \times 75 = \frac{950}{16} = 59.375\,DA$$

2.

c.  The probability that the destinations of the **03** passengers are different is none other than the number of Arrangements or combinations (without repetition) of 01 among $n_i$ divided by the number of combinations of 03 among 16.

Number of groups of 03 passengers: $C_{16}^3 = \dfrac{16!}{3! \times 13!} = 560$ .

Number of groups of 03 passengers to *different destinations*:

$$A_7^1 \times A_5^1 \times A_4^1 = C_7^1 \times C_5^1 \times c_4^1 = 7 \times 5 \times 4 = 140$$

Thus:
$$p = \frac{140}{560} = \frac{1}{4}$$

d.  At least *one* of the 03 passengers goes to station **B**:

$$p = \frac{C_7^1 \times C_9^2 + C_7^2 \times C_9^1 + C_7^3}{C_{16}^3} = \frac{7 \times 9 \times 4 + 7 \times 3 \times 9 + 7 \times 5}{560} = 0.85$$

Or, otherwise:
$$p = 1 - p(\bar{B}) = 1 - \frac{C_9^3}{C_{16}^3} = 1 - \frac{3 \times 4 \times 7}{560} = 0.85$$

e.  In this question, we specified in advance that the 03 passengers are heading together to the same station (**A** or **B** or **C**), and we are trying to find out the probability of their orientation towards **B**. So, we are looking for: $p_{3\,sets}(B)$.

So you have to determine $p(3\,sets)$ then $p_{3\,sets}(B)$. This is a case of Bayes' theorem.

$$p(3\,sets) = p(A3) + p(B3) + p(C3) = \frac{C_7^3}{C_{16}^3} + \frac{C_5^3}{C_{16}^3} + \frac{C_4^3}{C_{16}^3} = \frac{7}{80}$$

$$p_{3\,sets}(B) = \frac{p(3\,sets \cap B)}{p(3\,sets)} = \frac{C_7^3/C_{16}^3}{p(3\,sets)} = \frac{5/80}{7/80} = \frac{5}{7}$$

**Series 6:** Distribution laws

## Exercise 01

We have the binomial law: $p(X = k) = C_n^k \times p^k \times q^{n-k}$

*For* $n = 20$ ; $p = 0.4$ ; $q = 1 - p = 0.6$ ; *We obtain*: $p(X = k) = C_{20}^k \times 0.4^k \times 0.6^{20-k}$

It suffices to replace the value of *k* to find the desired probability.

$$p(X = 3) = C_{20}^3 \times 0.4^3 \times 0.6^{20-3} = \frac{20!}{3! \times (20-3)!} \times 0.4^3 \times 0.6^{17} \simeq 0.01235$$

$$p(X = 17) = C_{20}^{17} \times 0.4^{17} \times 0.6^3 \simeq 0.00004$$

$$p(X = 10) = C_{20}^{10} \times 0.4^{10} \times 0.6^{10} \simeq 0.117$$

$$p(X \leq 1) = p(X = 0) + p(X = 1) = C_{20}^0 \times 0.4^0 \times 0.6^{20-0} + C_{20}^1 \times 0.4^1 \times 0.6^{20-1}$$

$$= \frac{20!}{0! \times (20-0)!} \times 0.4^0 \times 0.6^{20} + \frac{20!}{1! \times (20-1)!} \times 0.4^1 \times 0.6^{19}$$

$$\simeq 0.00003656 + 0.0004875 \simeq 0.00052$$

$$p(X \geq 18) = p(X = 18) + p(X = 19) + p(X = 20)$$

$$p(X \leq 15) = 1 - p(X > 15) = 1 - \left[ p(X = 16) + p(X = 17) + p(X = 18) + p(X = 19) + p(X = 20) \right]$$

$$p(X \leq 10) = 1 - p(X < 10)$$

## Exercise 02

1. Drilling leads to an *oil slick* with probability **0.1** or *not* with probability **0.9**. It is therefore indeed a *Bernoulli test* with *parameter* **0.1**. There are *two possible outcomes*.

2.

- The *boreholes* must be *independent*, so that **X** follows a *binomial law*.

-    $p(X \geq 1) = 1 - p(X = 0) = 1 - C_9^0 \times 0.1^0 \times 0.9^{9-0} = 1 - 0.9^9 \simeq 0.613$

## Exercise 03

This is indeed a *Bernoulli variable* (test), which follows a *binomial distribution* $\beta(1000, 5 \times 10^{-3})$.

-
-

$$p(X = 2) = C_{1000}^2 \times 0.005^2 \times 0.995^{998} \simeq 0.084$$
$$p(X \leq 2) = p(X = 0) + p(X = 1) + p(X = 2)$$

$$= 0.995^{1000} + C_{1000}^1 \times 0.005^1 \times 0.995^{999} + 0.084 \simeq 0.124$$

-

$$p(X \geq 2) = 1 - p(X < 2) = 1 - \left[ p(X = 0) + p(X = 1) \right] \simeq 0.960$$

## Exercise 04

The statement of the experiment, in addition to the indication that we are interested in the color white after each draw, makes the experiment of the launch a case of a Bernoulli' random variable, with a probability of success $p=7/15$. In addition, the five (5) independent launches well describe a random variable, which follows a binomial law $\beta\left(5, \dfrac{7}{15}\right)$.

So, we can write the binomial law of this experiment as follows:

$$p(X = k) = C_n^k \times p^k \times q^{n-k} \quad (q = 1 - p)$$

We have:           $n = 5 \; ; \; p = \dfrac{7}{15} \Rightarrow q = \dfrac{8}{15}$

$$p(X = k) = C_5^k \times \left(\frac{7}{15}\right)^k \times \left(\frac{8}{15}\right)^{n-k}$$

-    For *exactly* a **single** white ball after five launches, just replace **k** by 1 in the law:

$$p(X = 1) = C_5^1 \times \left(\frac{7}{15}\right)^1 \times \left(\frac{8}{15}\right)^4 = \frac{5!}{1! \times 4!} \times \frac{7}{15} \times \frac{4096}{50625} = \frac{143360}{759375}$$
$$p(X = 1) \simeq 0.1887$$

- For a *maximum* of **two** white balls after the five throws, **k** is *less than or equal to 2*:

$$p(X \leq 2) = p(X = 0) + p(X = 1) + p(X = 2)$$

$$p(X = 0) = C_5^0 \times \left(\frac{7}{15}\right)^0 \times \left(\frac{8}{15}\right)^5 = \left(\frac{8}{15}\right)^5 = \frac{32768}{759375}$$

$$p(X = 0) \approx 0.0431$$

$$p(X = 2) = C_5^2 \times \left(\frac{7}{15}\right)^2 \times \left(\frac{8}{15}\right)^3 = 10 \times \frac{49}{225} \times \frac{512}{3375} = \frac{250880}{759375}$$

$$p(X = 2) \approx 0.33$$

$$p(X \leq 2) = \frac{32768}{759375} + \frac{143360}{759375} + \frac{250880}{759375} = \frac{427008}{759375}$$

$$p(X \leq 2) \approx 0.562$$

- To draw the bar graph of the experiment, we calculate the probabilities taken by the random variable for each value of **k** (*six values* for our experiment). We have already calculated 3 values (**k**=0; 1 and 2). We will calculate the others:

$$p(X = 3) = 10 \times \frac{343}{3375} \times \frac{64}{225} = \frac{219520}{759375}$$

$$p(X = 3) \approx 0.289$$

$$p(X = 4) = 5 \times \frac{2401}{50625} \times \frac{8}{15} = \frac{96040}{759375}$$

$$p(X = 4) \approx 0.1265$$

$$p(X = 5) = \left(\frac{7}{15}\right)^5 = \frac{16807}{759375}$$

$$p(X = 5) \approx 0.0221$$

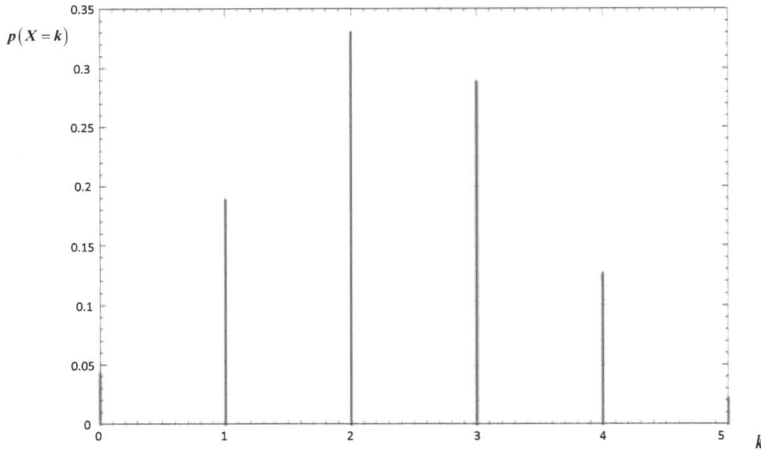

## *Calculation of* $E(X)$

We know that for a binomial law: $E(X) = n \times p$.

So:

$$E(X) = n \times p = 5 \times \frac{7}{15}$$
$$E(X) \simeq 2.334$$

# Series Correction Statistics

**Series 7**

## STATISTICS

### Exercise 01

We are asked to calculate for the series below, the *mean*, the *mode,* the *median* and the *extent*.

**Series 1:**    1   3   6   8   9   11   14   17   20   24   29

$$\circ \ \bar{X} = \frac{1}{N}\sum_{i=1}^{11} x_i = \frac{1+3+6+8+9+11+14+17+20+24+29}{11} \simeq 12.91 \ ;$$

o   $Mod = 0$   (No mode : all $x_i$ have the same frequency 1);

o   $Med = 11$ (The value 11 divides the series exactly into two equal parts. We have an odd number of values);

o   $Ext = x_{i\,max} - x_{i\,min} = 29 - 1 = 28$

**Series 2**

| $x_i$ | 02 | 04 | 07 | 14.5 | 10.0 | 8.8 | 07 | 09 | 03 |
|-------|----|----|----|------|------|-----|----|----|----|
| $n_i$ | 02 | 06 | 01 | 01 | 07 | 05 | 02 | 04 | 03 |

$$\circ \quad \bar{X} = \frac{1}{\sum_{i=1}^{9} n_i}\sum_{i=1}^{9} n_i \times x_i = \frac{2\times 2 + 6\times 4 + 1\times 7 + 1\times 14.5 + 7\times 10 + 5\times 8.8 + 2\times 7 + 4\times 9 + 3\times 3}{2+6+1+1+7+5+2+4+3}$$

$$= \frac{222.5}{31} \simeq 7.18$$

The series should be ordered before calculating the remaining parameters. We get the new representation:

| $x_i$ | 02 | 03 | 04 | 07 | 8.8 | 9 | 10 | 14.5 |
|---|---|---|---|---|---|---|---|---|
| $n_i$ | 02 | 03 | 06 | 03 | 05 | 04 | 07 | 01 |

- $Mod = 07$   ($x_i = 10$ has the greater frequency);

- $Med = \dfrac{7 + 8.8}{2} = 7.9$  (We have an even number of $x_i$, we take the mean value of the two middle ones);

- $Ext = x_{i\,max} - x_{i\,min} = 14.5 - 2 = 12.5$

## Exercise 02

We have the following statistics:

| Class | [A-B] | [B-C] | [C-D] | [D-E] |
|---|---|---|---|---|
| Center of the class | 7.5 | 12.5 | 17.5 | 22.5 |
| Workforce (Frequency) | 15 | 10 | 06 | 09 |

**1.**

- *Population*: The workers
- *Statistical variable*: The time
- *Its Type*: Quantitative (counted in minutes and therefore quantified by numbers)

**2.**

The lengths of the classes are the same $\Rightarrow$ **(B-A) = (C-B)** $\Rightarrow C = 2 \times B - A$

And   we   already   have:   **(B+A)/2=7.5**          and          **(C+B)/2=12.5**
$\Rightarrow B+A=15 \ \ and \ \ C+B=25$

$$\Rightarrow \ B=15-A \quad et \ \ 2\times B-A+B=3\times B-A=25$$

$$\Rightarrow \ 3\times(15-A)-A=25 \ \ \Rightarrow \ \ 45-4\times A=25 \ \ \Rightarrow \ \boxed{A=5}$$

$$B=15-A \ \Rightarrow \ \boxed{B=10}$$

$$similarly, \ we \ find: \boxed{C=15} \ ; \ \boxed{D=20} \ ; \ \boxed{E=25}$$

So, each class length equals: **05mn**

**3.**

$$\overline{X}=\frac{1}{\sum\limits_{i=1}^{5}n_i}\sum\limits_{i=1}^{5}n_i.x_i=\frac{1}{40}\left(15\times7.5+10\times12.5+6\times17.5+9\times22.5\right)=13.625\,mn$$

$$\sigma=\frac{1}{\sum\limits_{i=1}^{5}n_i}\sum\limits_{i=1}^{5}n_i.x_i^2-\overline{X}^2=\frac{1}{40}\left(15\times(7.5)^2+10\times(12.5)^2+6\times(17.5)^2+9\times(22.5)^2\right)-(13.625)^2$$

$$\simeq5.86\,mn$$

$$CV=\frac{\overline{X}}{\sigma}\simeq0.43<0.5$$

The value of **CV** is less than **0.5**, it means that the series is weakly dispersed. But, its value is ***close to 0.5*** is it anyway dispersed a considerable degree.

**4.**

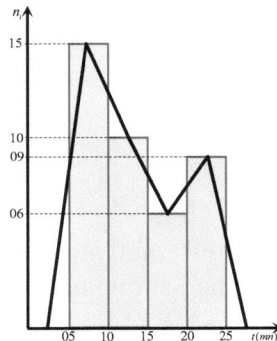

## Exercise 03

A quality control inspector extracted a 40-week sample from his database, where he noted $X$, the number of work accidents recorded per week. He obtained the following results:

2  0 4 2 2  1  3  2 0 5 4 3 2 4 5 6 6 4 2 0
3  4 4 2 6 2 4 3 0 4 3 4 3 3 5 5 4 2 2 1

We can therefore draw up the following table of frequencies:

**Frequency table of the number of accidents per week**

| Number of accidents/week | Absolute frequencies | Relative frequencies | Cumulative Relative Frequencies |
|:---:|:---:|:---:|:---:|
| 0 | 4 | 0.100 | 0.100 |
| 1 | 2 | 0.050 | 0.150 |
| 2 | 10 | 0.250 | 0.400 |
| 3 | 7 | 0.175 | 0.575 |
| 4 | 10 | 0.250 | 0.825 |
| 5 | 4 | 0.100 | 0.925 |
| 6 | 3 | 0.075 | 1.000 |
| *Total* | *n=40* | 1.000 | - |

As for the bar chart, we get something like:

**Note:** The bars must not be thick, because the variable takes exactly the values 0, 1, 2, ... You can add the numbers or the relative frequencies on the bars.

## Exercise 04

Let $X$ be the daily receipts (in dollars) of a small store. A sample of size $n=40$ days was randomly selected which yielded the following results:

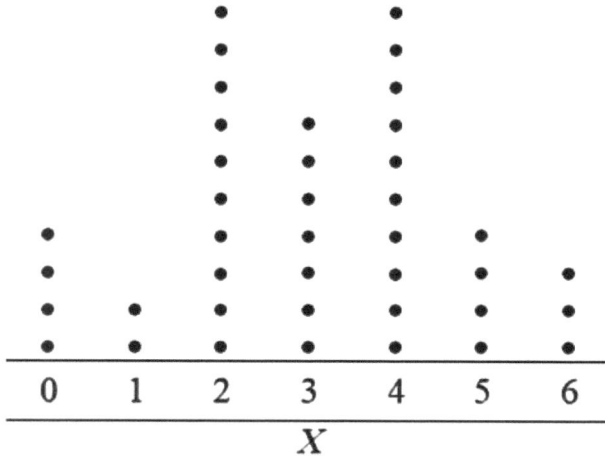

**Distribution of accidents over the weeks**

16.00　58.50　68.20　78.00　79.45　142.20　145.3　186.70　209.05　216.75

219.70　247.75　249.10　256.00　257.15　262.35　268.60　269.60　270.15　284.45

319.00　332.00　343.29　350.75　354.90　372.60　383.20　389.20　404.55　420.20

428.50　432.40　444.60　446.80　456.10　458.10　493.95　511.95　521.05　621.35

The number of classes to be trained is: $K = 1 + \dfrac{10}{3}\log(40) = 6.34 \approx 7\,classes$ .

of amplitude each equal to: $A = \dfrac{621.35 - 16}{7} = 86.48 \approx 90\,Dollars$ .

This amplitude is rounded to **90**. This gives the following table of frequencies, where the classes are intervals closed on the left and open on the right except the last which is an interval closed on both sides.

| Receipts($X$) | Absolute Frequencies | Relative Frequencies | Cumulative Relative Frequencies |
|---|---|---|---|
| [10 ; 100] | 5 | 0.125 | 0.125 |
| [100 ;190] | 3 | 0.075 | 0.200 |
| [190 ;280] | 11 | 0.275 | 0.475 |

*cont.....*

| | | | |
|---|---|---|---|
| [280 ;370] | 6 | 0.150 | 0.625 |
| [370 ;460] | 11 | 0.275 | 0.900 |
| [460 ;550] | 3 | 0.075 | 0.975 |
| [550 ;640] | 1 | 0.025 | 1.000 |
| **Total** | $n=40$ | 1.000 | |

- Below, The representative histogram is drawn, it is a series of rectangles juxtaposed to each other drawn above each of the classes, whose width is equal to the amplitude of the class (taken as a unit of measurement), and whose surface reflects the frequency relative of the class it represents.

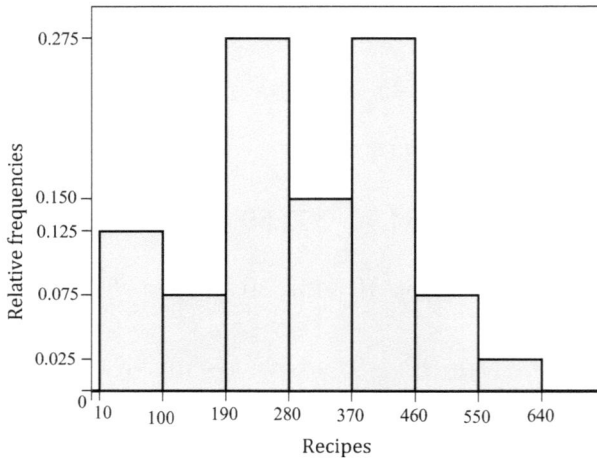

## Exercise 05

we are given the distribution of the continuous random variable provided with its frequencies.

| $x_i$ | [1-1.5] | [1.5-2] | [2-2.5] | [2.5-3] | [3-3.5] | [3.5-4] |
|---|---|---|---|---|---|---|
| $n_i$ | 10 | 15 | 26 | 31 | 10 | 08 |

- **Calculation of the arthmetic mean** $\overline{X}$: We can do this directly with the absolute frequencies ($n_i$) or by using the relative frequencies ($f_i$).

Since the values of $x_i$ are given on classes (intervals), we must take the center of the class $c_i$ as reference value (review the course).

For this exercise:

| $c_i$ | 1.25 | 1.75 | 2.25 | 2.75 | 3.25 | 3.75 |
|-------|------|------|------|------|------|------|

- *Direct calculation* (*with absolute frequencies*):

$$\bar{X} = \frac{\sum_{i=1}^{6} n_i \times c_i}{\sum_{i=1}^{6} n_i} = \frac{10 \times 1.25 + 15 \times 1.75 + 26 \times 2.25 + 31 \times 2.75 + 10 \times 3.25 + 8 \times 3.75}{10 + 15 + 26 + 31 + 10 + 8} = \frac{245}{100}$$

$$\bar{X} = 2.45$$

- *Calculation with absolute frequencies*:

For this exercise:      $$f_i = \frac{n_i}{\sum_{i=1}^{6} n_i} = \frac{n_i}{N} = \frac{n_i}{100}$$

| $f_i$ | 0.10 | 0.15 | 0.26 | 0.31 | 0.10 | 0.08 |
|-------|------|------|------|------|------|------|

$$\bar{X} = f_i \times c_i = 0.10 \times 1.25 + 0.15 \times 1.75 + 0.26 \times 2.25 + 0.31 \times 2.75 + 0.10 \times 3.25 + 0.08 \times 3.75$$
$$\bar{X} = 2.45$$

- **Calculation of the Mode**: The continuous nature of the random variable gives the modal class corresponding to the greatest frequency ($n_i$).

So:  **[2.5-3]**  $\rightarrow$  **31**

For the mode, we use the expression:        $$Mod = A_1 + \frac{\Delta_1}{\Delta_1 + \Delta_2} L_{Mod}$$

$A_1$ : Lower limit of the modal class (= **2.5**)

$\Delta_1$ : The difference between the frequency of the modal class and previous class (= **31-26=05**)

$\Delta_2$ : The difference between the frequency of the modal class and next class (= **31-10=21**)

$L_{Mod}$ : Length of the modal class (= **3.0-2.5= 0.5**)

$$Mod = 2.5 + \frac{5}{5+21} 0.5 \approx 2.59$$

From the expression for calculating the mode, we can see that it is the ***sum of*** the *minimum value of the modal class* and a *small averaged value*, evaluated by taking into consideration the frequencies of the classes before and after (when the series is ordered of course)

## REMARKS

o Remember that the statistical series must be ordered (increasingly or decreasingly) before proceeding with the calculation;

o The value of the *mode* must necessarily be inside the modal class. In addition, it can differ from the value of the center of the class unlike the case of a discontinuous random variable.

- **Calculation of the Median**: For its calculation, we must determine the *median class* which corresponds to the class with an *increasing cumulative frequency* $(ICF) \geq N/2$ ( $N = \sum_{i=1}^{6} n_i$ ).

For this exercise: *ICF*

| *ICF* | 10 | 25 | 51 | 82 | 92 | 100 |
|-------|-----|-----|-----|-----|-----|------|

So, the median class is **[2-2.5]**        (N/2=50)

We use the expression:        $Med = A_1 + \dfrac{N/2 - N_{Med-1}}{n_{Med}} L_{Med}$

$A_1$ : Lower limit of the median class (= **2**)

$N_{Med-1}$ : ACF of the class just before the median class (= **25**)

$n_{Med}$ : Frequency of the median class (= **26**)

$L_{Med}$ : Length of the median class (= **2.5-2= 0.5**)

$$Med = 2 + \frac{50-25}{26}0.5 \simeq 2.48$$

The median is the value of $x_i$ that divides the population into two equal halves. We can say that **50%** of the population has an $x_i$< 2.48, and the same for $x_i$> 2.48.

-   **Calculation of the first Quartile**: It correponds to the class with an *assanding cumulative frequency (ICF)* $\geq$ *N/4*.

Thus: $ICF \geq$ **25** and the class is **[1.5-2]**

We use the expression:    $Q_1 = A_{Q1} + \dfrac{N/2 - N_{Q1-1}}{n_{Q1}} L_{Q1}$

$A_1$ : Lower limit of the first quartile class (= **1.5**)

$N_{Q1-1}$ : *ICF* of the class just before the first quartile class (= **10**)

$n_{Q1}$ : Frequency of the first quartile class (= **15**)

$L_{Q1}$ : Length of the first quartile class (= **2-1.5= 0.5**)

$$Q_1 = 1.5 + \frac{25-10}{15}0.5 = 2.0$$

The first quartile is the value of $x_i$ corresponding to 25% of the population. In other words, 25% of the population have an $x_i$ less than $Q_1$ while 75% have a greater value.

# CONCLUSION

At the end of this first version of the book, I hope that the reader (student, teacher or other) has benefited from the content without difficulty. I repeat here that I presented the part on probabilities first because of its difficulty compared to that of statistics among students according to my experience teaching the module. In addition, probability calculations are essential for technical specialties more than economic ones where the part of statistics is the most encountered. I tried to detail in a chained way the theory, the process of the calculation and the exploitation of the definitions by discussed examples in order to show why we study this chapter or the other. I showed that the laws of probability are those of statistics by difference of domains only, where for the probabilities we make anticipations even of the problems to be studied themselves. *i.e*, we study the problem mentally and we imagine all its outcomes based on their probabilities, then we fix our decisions depending on the weighting or the rarity of the outcome sought (favorable case). On the other hand, in statistics, the problem and its outcomes already existed and the weightings are determined from the samples which change form and type according to the problem and the objectives. Therefore, mastering one part serves with excellence to master the other, despite the recommendation to start with the part of probabilities boosted by the factor of imagination. I hope that in future versions, enrichment in quality and content will soon be assured.

# REFERENCES

[1]    C. Ash, "*The Probability Tutoring Book: An Intuitive Course for Engineers and Scientists (and Everyone Else!)* ". Wiley-IEEE Press: New York, 1993.

[2]    R. Durrett, Probability: "Theory and Examples. 2ed ed, Duxbury Press: California", Available from: R. Durrett

[3]    D. Bertsekas, J.N. Tsitsiklis, "Introduction to Probability". 2ed ed, Athena Scientific: Massachusetts, J.N. Tsitsiklis Available from: D. Bertsekas

[4]    C.M. Grinstead, and J.L. Snell, *Introduction to Probability.* University Press of Florida: Florida, 2009.

[5]    H. Tijms, "Understanding Probability: Chance Rules in Everyday Life. 2ed ed, Cambridge University Press: Cambridge", Available from: H. Tijms

[6]    H.T. Nguyen, and T. Wang, "A graduate course in probability and statistics, Volume I: Essentials of probability for statistics", In: *Essentials of probability for statistics.* Tsinghua University Press: Beijing, 2007.

[7]    R.B. Ash, *Basic Probability Theory.* Dover Publications: New York, 2012.

[8]    M.G. Bulmer, *Principles of Statistics.* M.I.T. Press: Cambridge, Massachusetts, 1967.

[9]    L. Wasserman, and L.A. Wasserman, *All of Statistics: A Concise Course in Statistical Inference.* Springer Science & Business Media: Berlin, Heidelberg, 2004. http://dx.doi.org/10.1007/978-0-387-21736-9

[10]   D. Griffiths, "Head First Statistics", In: *O'Reilly Medea.* O'Reilly Medea, 2008.

[11]   H.T. Nguyen, and T. Wang, *A graduate course in probability and statistics, Volume I: Essentials of probability for statistics.* Tsinghua University Press: Beijing, 2007.

[12]   Triola, Mario F., William Martin Goodman, Richard Law, and Gerry Labute. Elementary statistics. Reading, MA: Pearson/Addison-Wesley, 2006.

[13]   D.S. Shafer, "Introductory statistics", *Flat World Knowledge,* New York, 2013.Z. Zhang

[14]   K.M. Ramachandran, and C.P. Tsokos, *Mathematical Statistics with Applications.* Elsevier Academic Press: Cambridge, Massachusetts, 2009.

# SUBJECT INDEX